Analog Circuits and Signal Processing

AF618894

Series Editors
Mohammed Ismail
The Ohio State University Dept. Electrical & Computer Engineering
Dublin
Ohio
USA

Mohamad Sawan
École Polytechnique de Montréal
Montreal
Québec
Canada

More information about this series at http://www.springer.com/series/7381

Ying Cao • Paul Leroux • Michiel Steyaert

Radiation-Tolerant Delta-Sigma Time-to-Digital Converters

Ying Cao
KU Leuven
Heverlee
Belgium

Paul Leroux
KU Leuven
Heverlee
Belgium

Michiel Steyaert
Department of Electrical Engineering
KU Leuven
Heverlee
Belgium

ISSN 1872-082X ISSN 2197-1854 (electronic)
Analog Circuits and Signal Processing
ISBN 978-3-319-38529-7 ISBN 978-3-319-11842-0 (eBook)
DOI 10.1007/978-3-319-11842-0

Springer Cham Heidelberg New York Dordrecht London
© Springer International Publishing Switzerland 2015
Softcover reprint of the hardcover 1st edition 2015
This work is subject to copyright. All rights are reserved by the Publisher, whether the whole or part of the material is concerned, specifically the rights of translation, reprinting, reuse of illustrations, recitation, broadcasting, reproduction on microfilms or in any other physical way, and transmission or information storage and retrieval, electronic adaptation, computer software, or by similar or dissimilar methodology now known or hereafter developed.
The use of general descriptive names, registered names, trademarks, service marks, etc., in this publication does not imply, even in the absence of a specific statement, that such names are exempt from the relevant protective laws and regulations and therefore free for general use.
The publisher, the authors and the editors are safe to assume that the advice and information in this book are believed to be true and accurate at the date of publication. Neither the publisher nor the authors or the editors give a warranty, express or implied, with respect to the material contained herein or for any errors or omissions that may have been made.

Printed on acid-free paper

Springer is part of Springer Science+Business Media (www.springer.com)

To Peizi. For Everything. Forever.
And to Alex and Erik.

Preface

This book, based on pioneering PhD work at KU Leuven (University of Leuven), Belgium, offers a truly unique combination of techniques and insights in time-domain analog signal processing and radiation hardening by design methodologies.

With the continued and progressive scaling of Complementary metal–oxide–semiconductor (CMOS) chip technologies, a change in paradigm has been established where the conversion of analog to digital signals is shifted to the time domain. Indeed, with the transistor dimensions also the supply voltage becomes lower which tends to limit the dynamic range of classic analog-to-digital converters (ADCs). On the other hand, the improved high-frequency performance in nanometer technologies enables higher resolution in the time domain where the analog information is stored in the time difference between the zero-crossings of an input signal. The ADC is replaced by a time-to-digital converter (TDC).

The need for high-resolution TDCs is numerous as they can be used for all-digital phase-locked loops (ADPLLs), communication circuits, positioning, sensor readout in instrumentation, time-of-flight measurement, digital oscilloscopes, particle physics, and many more. Different TDC topologies, explained in this book, have been proposed, inspired by the corresponding family of ADCs. Nevertheless, the one family of $\Delta\Sigma$ ADCs had stayed without TDC counterpart for many years owing to the intrinsic problem of storing and integrating time. Only recently the Holy Grail of higher order noise-shaping $\Delta\Sigma$ TDCs was unveiled by the authors. The corresponding groundbreaking circuit architectures form the highest motivation of this book. $\Delta\Sigma$ TDCs are of unprecedented importance since most applications require an extreme temporal resolution down to picoseconds while the analog time input intrinsically varies at a much lower rate (microseconds).

It does not stop here, as the book even takes it one massive leap further. The circuits described in this work have been designed to survive the harshest radiation environments that are found in the most challenging nuclear physics projects in history, namely, the International Thermonuclear Experimental fusion Reactor (ITER) and the Large Hadron Collider (LHC) at CERN. The book discusses all layout-level, circuit-level, and architectural measures needed to design MegaGray (MGy) radiation tolerant integrated circuits (ICs) in nanometer CMOS technologies. This also makes the circuits inherently ultra-robust to process variability, temperature, and aging effects and paves the way for a larger penetration of advanced integrated electronics in the nuclear and space sector.

This book on radiation tolerant $\Delta\Sigma$ TDCs is considered indispensable for all engineers, circuit designers, and postgraduate students who are engaged in the design of ADPLLs, TDCs, or other time- or phase-based signal processing circuits. The presented circuits and techniques are essential for all engineers involved in radiation hardened IC design or harsh environment electronics. The book is instructive and informative for all electronic engineers wanting to know more on the design of time-based signal processing circuits or radiation hardening by design.

Enjoy reading!

Heverlee
October 2014

Paul Leroux
Ying Cao

Contents

Abbreviations and Symbols

ADC	Analog-to-digital converter
CTAT	Complementary to absolute temperature
DBLC	Dynamic base leakage compensation
DR	Dynamic range
DTMOS	Dynamic-threshold-voltage MOSFET
ELT	Enclosed layout transistor
ENOB	Effective number of bits
FOM	Figure-of-merit
GRO	Gated ring oscillator
IEF	Integrated error feedback
ITER	International Thermonuclear Experimental Reactor
LBE	Lead–Bismuth eutectic
LHC	Large Hadron Collider
LIDAR	Light detection and ranging
LPI	Local passive interpolation
MASH	Multistage noise-shaping
MYRRHA	Multipurpose hybrid research reactor for high-tech applications
OSR	Oversampling ratio
PAF	Power averaging feedback
PLL	Phase-locked loop
PN	Phase noise
PTAT	Propotional to absolute temperature
PWM	Pulse-width modulated
PVT	Process, voltage, and temperature
RHBD	Radiation-hardened-by-design
RHBP	Radiation-hardened-by-process
SAR	Successive approximation
SC	Switched capacitor
SCR	Semiconductor-controlled rectifier
SEE	Single-event effect
SEL	Single-event latchup
SET	Single-event transient

SEU	Single-event upset
SNACC	Sensitive node active charge cancellation
SNDR	Signal-to-noise plus distortion ratio
STI	Shallow trench isolation
TA	Time amplifier
TC	Temperature coefficient
TDC	Time-to-digital converter
TIA	Transimpedance amplifier
TID	Total ionizing dose
TOF	Time of flight
TVC	Time-to-voltage conversion
VCSEL	Vertical-cavity surface-emitting laser
VCO	Voltage-controlled oscillator
VDC	Voltage-to-digital conversion
WSN	Wireless sensor nodes

Symbols

α	reference voltage to supply ratio
ΔV_{it}	MOS threshold voltage shift due to radiation-induced interface trap
ΔV_{ot}	MOS threshold voltage shift due to radiation-induced oxidetrap
e_{skew}	phase skew error
F_{s}	raw resolution of a TDC
f_{m}	carrier offset frequency
G_y	gray, a unit of absorbed radiation dose
I_B	diode's base current
I_C	diode's collector current
I_{ref}	reference current
L	order of noise-shaping function
$L(f_m)$	phase noise
N_{ot}	total number of radiation induced holes in MOS gate oxide
P_{diss}	dissipated power
q_{err}	quantization error
rad	a unit of absorbed radiation dose, 1 rad = 0.01 Gy
Sv	sievert, a unit of absorbed dose by human body
$\sigma_{\Delta T_{osc}}$	jitter variance
T_{d}	delay time of a CMOS delay cell
T_{eff}	effective resolution of a TDC
T_{fs}	full-scale input level
T_{OSC}	raw resolution of a TDC
t_{cmp}	comparator delay
t_{ox}	MOS gate oxide thickness
V_{EB}	emitter-base voltage
V_{ref}	reference voltage
V_{dd}	power supply voltage

List of Figures

List of Tables

Chapter 1
Introduction

Abstract In recent years, the radiation hardness of integrated circuits has drawn more and more attention due to their increasingly important role in electronic systems in space, nuclear, and high-energy particle physics applications. Comprehensive background information regarding this new but rather crucial feature of integrated circuits are discussed in this chapter. Starting with an inspiring story about how integrated circuits pave the way to the Higgs particles, some most interesting potential applications of radiation-hardened electronics are then introduced. It is followed by a detailed explanation of a special implementation, which is the main topic of this book: a radiation-hardened time-to-digital converter for light detection and ranging.

1.1 Tiny Chips, Big Physics

Integrated circuits (IC), with its history back to the 1950s, have been continuously redefining the functionalities of electronics and reshaping people's way of living. Today, ICs are used in all electronic equipment and have revolutionized the world of electronics. Computers, mobile phones, and other digital consumer electronics are now available on most people's desk or in their pockets. Apart from these well-known stories, ICs have also played crucial roles in many other industries and subjects, for instance, particle physics.

With the recent discovery of the Higgs boson (tentatively confirmed to exist on 14 March 2013), the European Organization for Nuclear Research (CERN) has again stood under the spotlight. The search for the Higgs boson is carried out from the Large Hadron Collider (LHC) in Geneva, Switzerland, the largest particle accelerator in the world. Two independent experimental teams (the Compact Muon Solenoid (CMS) and the A Toroidal LHC Apparatus (ATLAS) team) both observed the existence of the Higgs boson based on a statistical analysis of numerous data and images produced from their detectors. The two detectors have similar structures and work on the same principle.

Just on the innermost tracking layer of the ATLAS detector, there are 1456 silicon pixel modules. Each module is 62.4-mm long and 21.4-mm wide, with 46,080 pixel elements read out by 16 chips, each serving an array of 18×160 pixels. The readout

© Springer International Publishing Switzerland 2015
Y. Cao et al., *Radiation-Tolerant Delta-Sigma Time-to-Digital Converters*,
Analog Circuits and Signal Processing, DOI 10.1007/978-3-319-11842-0_1

chips must withstand more than 300 kGy[1] of ionizing radiation and more than 5×10^{14} neutrons per cm^2 more than 10 years of operation [80]. ICs have also been used in the semiconductor tracker (SCT), which contributes to the measurement of momentum, impact parameter, and vertex position of charged particles. There are more than 4000 individual modules in the SCT, and 12 readout application-specific integrated circuits (ASICs) per module. Each IC covers 128 strips for a total of about 6.3 million readout channels (one channel per strip) [62]. One could imagine, without the implementation of radiation tolerant integrated sensor readout circuits, the ATLAS detector as well as the CMS detector cannot be built in such a compact way or even would never be constructed.

Higgs Boson and the ATLAS Detector

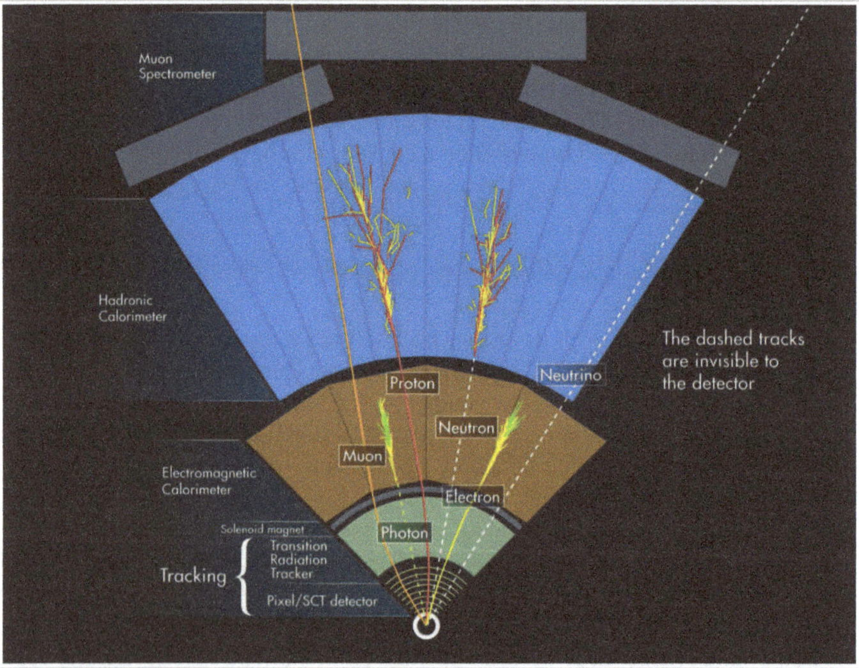

Fig 1.1 Illustrative cross section of the ATLAS detector. (Image source: CERN).

To produce Higgs bosons, two beams of particles are accelerated to high energies and allowed to collide within the LHC. Occasionally, a Higgs boson will be created as part of the collision by-products. Because the Higgs boson

[1] Gy, the *gray*, is the SI derived unit of absorbed dose. It is defined as the absorption of 1 J of such energy by 1 kg of matter. 1 Gy = 100 rad. When X-rays and gamma rays are applied on human body, another unit *sievert* (symbol: Sv) is often used. 1 Gy = 1 Sv. The radiation dose used in a single full-body CT scan is typically 10–30 mSv.

decays very quickly, particle detectors cannot detect it directly. Instead, the detectors register all the decay products (the decay "signature") and from this data the decay process is reconstructed. If the observed decay products match a possible decay process of a Higgs boson, this indicates that a Higgs boson might have been created.

Figure 1.1 shows the cross section of the ATLAS detector [19]. It consists of multiple layers, and each layer exploits the different properties of particles to catch and measure the energy or momentum of each one. A particle emerging from the collision and traveling outwards will first encounter the tracking system, made of silicon pixel detectors. The momentum of the particle can be calculated out by tracking its path through a magnetic field: The more curved the path, the less momentum the particle had.

1.2 Integrated Circuits for Space and Nuclear Instrumentation

1.2.1 Microelectronic Circuits for Space Missions

Besides their application in high-energy particle physics experimental machines, ICs have also been widely used in space and nuclear instrumentation. For space physics instruments, mainly equipped on satellites and space stations, various detectors and sensor interfaces are required to measure basic physical parameters such as magnetic field, radiation level, particle fluxes, energies, etc. Microelectronic circuits are key driving forces in reducing the total mass of space vehicles and enhancing their reliability. Due to the high radiation level in space, all semiconductor devices used for space instrumentation have to be radiation hardened. There are three main sources of radiation in space [29], which are:

- The trapped electrons and protons in the Earth's radiation belts.
- The protons and heavy ions produced by the solar particle events (SPE).
- The galactic cosmic rays (GCR) protons and heavy ions.

The radiation hardness requirements of electronic systems vary among different classes of space missions. For example, global monitoring for Earth and security and Earth observation missions are normally in low Earth orbit, thus well inside magnetosphere, which deflects most charged particles carried by solar winds and cosmic rays. Therefore, their accumulated doses are quite low, which is only 1–5 krad/year.

However, for some deep space missions, such as the future European Space Agency (ESA) flagship Jupiter science mission, the satellite will face a much harsher radiation environment. It is planned to be launched in 2022, arriving at Jupiter in 2030 to spend at least 3 years making detailed observation with multiple science payloads of the planets and its inner Galileian moons. Once in Jovian orbit, accumulated dose is foreseen to be hundreds of kiloradiations per year, where several highly variable trapped belts of protons, electrons, and plasma exist. Onboard electronic components are thus required to withstand at least 1 Mrad total ionizing dose (TID) level. Such missions are clearly a once-in-a-lifetime scientific opportunity and any failure at platform and payload level is considered not acceptable. These strict reliability, mass, power performance requirements are difficult to meet with current commercial available components and systems with sufficient radiation tolerance margins. Further investigation and improvement of ICs' radiation tolerance are quite necessary and urgent.

1.2.2 Electronics and Radiation Hardening in the Nuclear Industry

A similar story happened also in the nuclear industry. In conventional fission reactors, it is impossible or dangerous for human beings to conduct inspections or repair any components in the reactor vessel, where the radiation level compromise the safety of the personnel. Therefore, remote handling robots are needed to execute those hazardous tasks. Radiation-hardened instrumentation electronics (e.g., force/torque sensor, temperature sensor, image sensor, etc.) are thus highly demanded in these applications.

1.2.2.1 Handling of Nuclear Waste

An example of robotic arms replacing the human arm in the nuclear industry is the handling and processing of nuclear waste. Nowadays, in advanced nuclear facilities, the manipulation of hazardous material is often executed by a tele-operated master/slave system [36]. In such a system, the movement of the operator is reproduced by the slave robot through electrical control signals. If sensors are measuring the displacement of a master arm, then the information is sent to actuators on the slave robot. Since the cost of custom-built robot is often excessive, the strategy is to use commercially available robots. For radioactive environments, the requirements for a tele-operated robot are a good sealing of the parts to avoid contamination, the installation of a force feedback system, an acceptable level of radiation hardness, a high reliability, and an easy integration into embedded equipment. Since the manipulator is used in place of a human arm, it is not sufficient to simply visualize an operation to be able to operate correctly. The installation of a radiation-hardened force feedback system is very important when upgrading a commercial manipulator to its nuclear equivalent.

1.2.2.2 Decommissioning

In the recent decade, with many nuclear facilities reaching the end of their lifetimes, the decommissioning effort is a new priority. Decommissioning is part of the final shutdown of a nuclear reactor, which begins with the removal of highly radioactive spent fuel and may end with the cleanup of an entire facility or site, including in some cases contaminated soil and groundwater. Decommissioning involves the demolition of buildings and other structures, including the parts near the reactor core that may have become radioactive, as well as onsite handling of construction materials (mostly steel and concrete) and the packaging and transport of these materials for safe storage and disposal [39]. The reduction of human worker exposure as well as cost reduction requires the uses of robots for contaminated parts. The same tele-operated master/slave robotic system mentioned in the previous section can be used here to carry out these tasks.

WORLD REACTORS' STATUS STATISTICS

As of October 2013, 147 civilian nuclear power reactors have been shut down in 19 countries, including 32 in the USA, 29 in the United Kingdom, 27 in Germany, 12 in France, and 9 in Japan. The typical design life of a civilian nuclear power reactor is 30–40 years. There are currently 437 such reactors in operation worldwide. Of these reactors, 198 are more than 30 years old and 45 are more than 40 years old [39]. The backlog of civilian nuclear reactors that have been shut down but not yet decommissioned is expected to grow.

1.2.2.3 Nuclear Accident Intervention

When an unexpected problem occurs in a radioactive environment, it becomes a potential hazard. A fast and versatile tool for situation assessment is needed without putting personnel at risk. In March 2011, a tragic earthquake and subsequent tsunami caused the accident at the Fukushima nuclear power plant whereby radioactive material was released in the air and sea. Contamination of the reactor site and its surroundings made an area within a radius of 30 km uninhabitable.

Under these circumstances, remote-controlled robots are ideal candidates to conduct inspection in the nuclear facility, to monitor radiation level at the reactor site, or to shut down the troubled reactor. Unfortunately, even for a country like Japan, which is a superpower, with some of the world's most advanced robotic systems and the highest levels of industrial automation, it failed to send robots to fix the damaged reactors. One of the main reasons why a normal civil robot cannot complete this task is onboard electronics are not hardened enough against high-level radiation (the radiation dose rate at 1 cm from 2 g of spent fuel is in the order of 10 Gy/h). The radiation will damage microchips and sensors, and also corrupt data in silicon memories. Therefore, the development of nuclear industry-level robots equipped with megagray radiation tolerant electronics for a fast postaccident intervention becomes a very urgent mission.

1.2.2.4 Remote Handling in Fusion Reactors

Fusion power is the power generated by nuclear fusion processes. Fusion is the process at the core of the Sun. The most efficient fusion reaction to reproduce in the laboratory is the reaction between two hydrogen (H) isotopes deuterium (D) and tritium (T). The D–T fusion will produce one helium nuclei, one neutron, and energy. However, the fusion reaction requires a temperature of 150,000,000 °C to take place. At extreme temperatures, electrons are separated from nuclei and a gas becomes a plasma: A hot, electrically charged gas. Some of the key features of fusion make it an attractive option as part of a future energy mix. Fusion fuels are abundantly available and inherently safe. There is no possibility of a catastrophic accident in a fusion reactor resulting in major release of radioactivity to the environment or injury to non-staff, unlike modern fission reactors. The primary reason is that nuclear fusion requires precisely controlled temperature, pressure, and magnetic field parameters to generate net energy. If the reactor were damaged, these parameters would be disrupted and the heat generation in the reactor would rapidly cease [40].

One of the most-researched candidates for producing controlled thermonuclear fusion power is the tokamak. A tokamak is a device using a magnetic field to confine a plasma in the shape of a torus. The International Thermonuclear Experimental Reactor (ITER), the world's largest tokamak, is based on the "tokamak" concept of magnetic confinement, in which plasma is contained in a doughnut-shaped vacuum vessel. Strong magnetic fields are used to keep the plasma away from the walls; these are produced by superconducting coils surrounding the vessel, and by an electrical current driven through the plasma.

Remote handling will have an important role to play in the ITER tokamak. When operation begins, it will be impossible to make changes, conduct inspections, or repair any of the tokamak components in the activated areas other than by remote handling. Very reliable and robust remote handling techniques will be necessary to manipulate and exchange components weighing up to 50 t. The reliability of these techniques will also impact the length of the machine's shutdown phases [40]. All remote handling techniques developed for ITER operate on the same principle. A remote manipulator is used to detach the component. The component is then removed through a port and placed into the docked transport cask. The cask is moved on air bearings along to the hot cell facility. A similar docking occurs at the hot cell facility and the component is removed to be repaired or replaced. The process is then reversed to bring that component back to the vacuum vessel.

During the ITER nuclear phase, the decay gamma dose rates during in-vessel maintenance is up to several hundreds of grays per hour [68]. Therefore, all necessary electronic equipment such as force feedback systems, viewing systems, and communication systems used on those remote handled robotic vehicles and manipulators have to be radiation tolerant, due to the contaminated nature of the environment. For example, an in-vessel 3D viewing system, which is a fundamental tool to perform in-vessel inspections between plasma pulses or during a shutdown, needs to withstand ultra-high gamma dose rates and total doses up to 5 kGy/h and 5 MGy [68].

1.3 The MYRRHA Reactor

In recent years, interest has grown in the possibility of separating the long-lived radioactive waste from the used fuel and transmuting it into shorter-lived radionuclides so that the management and eventual disposal of this waste is easier and less expensive. The transmutation of long-lived radioactive waste can be carried out in an accelerator-driven system (ADS), where neutrons produced by an accelerator are directed at a blanket assembly containing the waste along with fissionable fuel. Following neutron capture, the heavy isotopes in the blanket assembly subsequently fission, producing energy in doing so.

A demonstration of the ADS technology can be found in the multipurpose hybrid research reactor for high-tech applications (MYRRHA), which is currently being developed at the Belgian nuclear research center, SCK·CEN, in Mol, Belgium [1]. The MYRRHA reactor, conceived as an ADS, is able to produce sustainable fission energy by bombarding the harmful nuclei with a high-neutron flux to transmute long-lived radioactive waste.

Structure of the MYRRHA Reactor

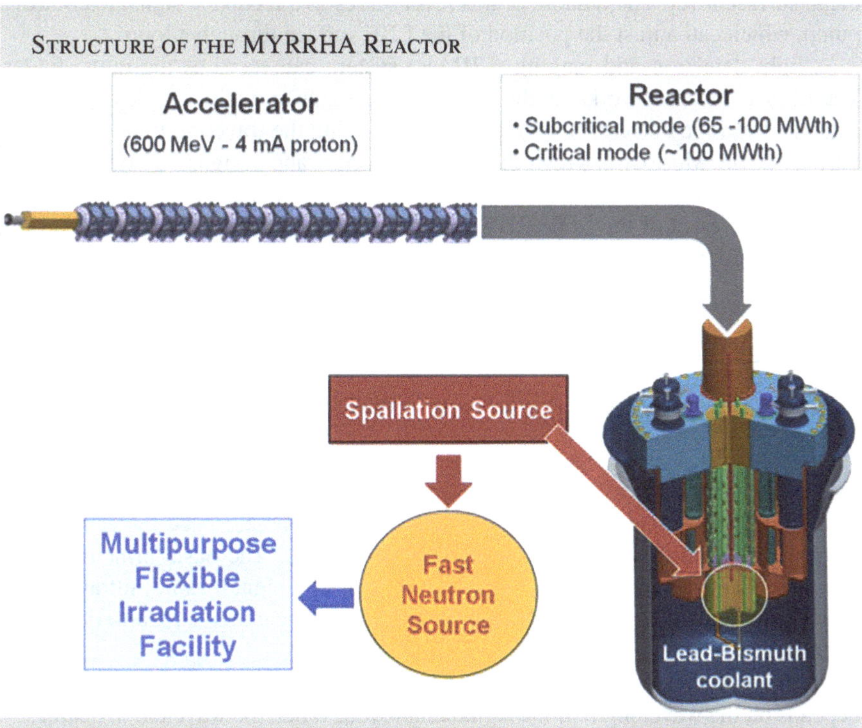

Fig 1.2 Demonstration of the MYRRHA reactor. (Image source: SCK·CEN)

Since neutrons cannot be directly accelerated, an ADS is actually a neutron source created by coupling a proton accelerator, a spallation source and a subcritical core [72]. A demonstration of the MYRRHA reactor is shown in Fig. 1.2. The accelerator is the driver of the ADS system. It provides the high-energy protons that are used in the spallation target to create neutrons which in their turn feed the subcritical core. This configuration gives a significant advantage of the ADS which is its inherent safety: the reactor is switched off at the moment the proton beam is switched off.

An ADS consists of three major parts: a proton accelerator, a spallation source, and a subcritical core. In the MYRRHA reactor, a liquid metal, for which lead-bismuth eutectic (LBE) has been chosen, is selected as the spallation source. In the spallation target, the position of the free liquid metal surface needs to be stabilized in order to guarantee proper heat exchanging and maintaining the spallation process. This can be done by first using a range-finding device to continuously monitor the liquid metal surface level. The readout of this level serves as a feedback signal to control a pump, which can adjust the position of the LBE surface through a loop.

A light detection and ranging (LIDAR) system can serve as the range finder. The LIDAR detector works on the time of flight (TOF) principle. It determines the distance from the detector to an object by measuring the traveling time of the light during the trip. More details regarding the architecture and design of a LIDAR system will be given in the next section. In our application, since the LIDAR system is placed in the inner vessel of the MYRRHA reactor , where enriched with gamma rays, it has to be hardened by design against radiation damage.

1.4 Light Detection and Ranging

1.4.1 LIDAR Techniques

LIDAR systems are widely used in a variety of remote sensing applications, e.g., positioning of vehicles, detecting particles in atmosphere, space target tracking, 3D imaging, and range finding in harsh environments. The basic principle of all active noncontact range-finding devices is to project a signal (radio, ultrasonic, or optical) onto an object and to process the reflected or scattered signal to determine the distance.

A common factor to all these range finder types is that the angular resolution depends on the wavelength of the signal employed. When the wavelength decreases, the divergence of the projected signal beam decreases. Therefore, if a high-resolution range finder is needed, an optical source must be chosen because radio and ultrasonic waves cannot be focused adequately [5]. Specifically, lasers are often used as the optical source due to their low divergence over long distance and monochromaticity. Optical distance measurement methods can technically be put into three categories: triangulation, time of flight, and interferometry methods [15].

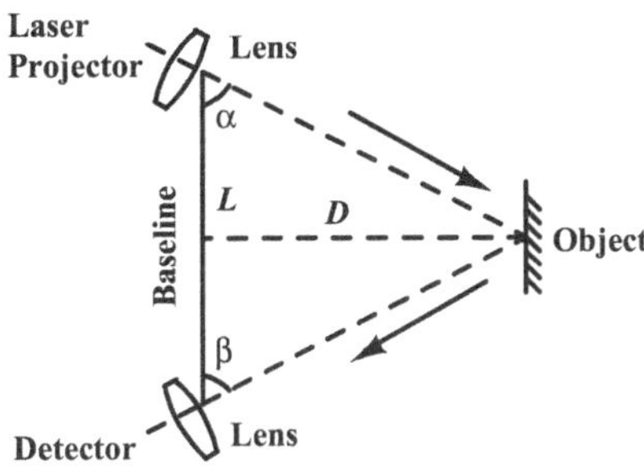

Fig. 1.3 Principle of triangulation detecting system

1.4.1.1 Triangulation

As illustrated in Fig. 1.3, a laser projector and a detector are allocated on the same baseline, where their position can be adjusted along the line. A laser signal is emitted from the projector. When it reaches an object, the reflected signal can be captured by a detector. By knowing the two angles α (between the baseline and the projector axis), β (between the baseline and the detector axis)[2], and the length of the baseline L, the distance from the detection system to the object can be calculated as:

$$D = \frac{L}{\cot(\alpha) + \cot(\beta)}. \tag{1.1}$$

An active device can be used here to perform photo sensing and signal processing, which can be built on complementary metal–oxide–semiconductor (CMOS) technology [35]. However, the triangulation method also requires sophisticated laser projecting system configuration, which involves many mechanical movements. This feature makes it unfavorable for applications in harsh environment like inside a nuclear reactor, where frequent turbulences are foreseen.

1.4.1.2 Time of Flight

The principle here is simple. The flight time of a light signal from a laser transmitter to the target and back to an optical detector, T, is measured, as shown in Fig. 1.4. The distance from the detector to the target can be determined based on the TOF theory. If the emitted light signal is a pulse, T is simply the time difference between the rising edges of the emitted signal and the reflected signal. The distance D is thus equal to $c \cdot T/2$, where c is the speed of the light.

It is also possible to measure the distance by using a linear frequency-modulated (FM) continuous-wave (CW) optical signal [5]. The similar TOF principle still applies here: by knowing the instantaneous frequency difference (f_b) between the

[2] When the object is in parallel with the baseline, $\alpha = \beta$.

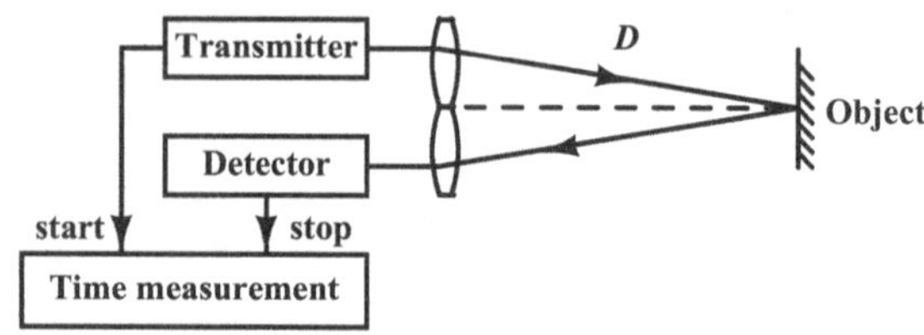

Fig. 1.4 Configuration of a time of flight (TOF) measurement system

output signal and the reflected signal, the distance from the detector to the object can be calculated as

$$D = \frac{1}{2} \cdot c \cdot \frac{f_b \cdot T_m}{\Delta f}, \tag{1.2}$$

where T_m denotes the FM ramp period (in case of a linear sawtooth frequency ramp), and Δf is the maximum frequency tuning range. Since the ramp period T_m can be chosen arbitrarily, the FMCW range finder can determine D values in the millimeter range. Unfortunately, the maximal measurable distance and resolution of a FMCW system is ultimately limited by the nonlinearity and phase noise of the FM laser diode. In general, the frequency of a laser diode versus its control current characteristic is nonlinear. As a consequence, deviations from the linear ramp usually occur, in turn, lead to a variation of the instantaneous frequency difference f_b. Furthermore, the phase noise of the laser diode also sets a limitation on its spectral linewidth, which also deteriorates the measurement accuracy. Experimentally, in order to achieve a millimeter distance accuracy, a laser diode with a frequency tuning range of several hundred gigahertz would be required. This imposes very stringent requirements on the performance of the CMOS optical receiver, which significantly increases the system's complexity, and it is usually impractical.

The TOF theory can be further inferred to the phase domain, where a sine-like CW laser signal is often used. For example, in a phase-shift range finder [65], the optical power is modulated with a constant frequency f_0. The propagation time of the light signal Δt is now converted into a phase-shift $\Delta\varphi = 2\pi f_0 \Delta t$. The distance information is then represented by the phase difference ($\Delta\varphi$) between the emitted and received signal, which is

$$D = \frac{1}{2} \cdot c \cdot \frac{\Delta\varphi}{2\pi f_0}. \tag{1.3}$$

The unambiguous distance measurement range of the CW phase-shift range finder is limited to $c/2f_0$, when $\Delta\varphi = 2\pi$. For instance, in order to obtain a unambiguous measurement range of larger than 100 m, a modulating frequency of lower than 1.5 MHz has to be used. However, this will limit its distance resolution when such a low modulating frequency is used, according to Eq. (1.3). Therefore, the CW phase-shift method is not suitable for applications when long measurement range and high resolution are required at the same time [43]. By contrast, there is no such trade-off in a pulsed TOF range finder. Its unambiguous detection range will only be limited by the repetition rate and output power of the laser transmitter. The spatial resolution depends mainly on the noise performance and time resolution of the receiver circuits. Moreover, the pulsed TOF range finder is inherently more robust than the CW range finder in a noisy environment since the accuracy of the latter highly relies on the amplitude of the reflected signal, which can be easily disturbed. Due to these reasons, the pulsed TOF range finder is a more suitable choice for the MYRRHA application. More details regarding the system design of a pulsed TOF laser range finder will be discussed in the next section.

1.4.1.3 Interferometry

An interferometer can be used for the accurate measurment of displacement and absolute distance. If a laser is FM, the time of flight Δt it takes for the light emitted from the laser diode to reach a target at a distance D and be scattered back leads to a frequency difference (Δf) between the emitted and backscattered light. If the functional dependence of the optical frequency f with respect to time t is linear, $\mathrm{d}f/\mathrm{d}t$ is constant and equal to a. When the two frequencies are interfered, an amplitude modulation is produced with a beat frequency given by

$$f_b = a \cdot \Delta t = a \cdot \frac{2D}{c}. \tag{1.4}$$

By measuring f_b, the distance can be calculated [13, 79]. The interferometry range finders are capable of providing excellent distance resolution with sub-*mm* accuracy. However, their measuring range can often not exceed 10 m due to the limited frequency tuning linearity of laser doides.

1.4.2 Pulsed TOF Laser Range Finder

A pulsed TOF laser range-finding device typically consists of the following: (1) a pulsed laser transmitter, (2) a receiver channel and (3) a time-to-digital converter (TDC) [70], as shown in Fig. 1.5. The operation of the device is based on measurement of the transit time of a short laser pulse from the transmitter to the optically visible target and back to the receiver. The laser transmitter, for instance, a vertical-cavity surface-emitting laser (VCSEL), produces a short optical pulse. Simultaneously, an optical reference signal is taken directly to a photodetector, where it is converted to an electrical pulse. This pulse is then amplified to generate a logic-level time reference for the time interval measurement unit (e.g., a TDC) as a start pulse. In the same way, a stop pulse is generated from the reflected light signal. The TDC compares the time difference between the start pulse and the stop pulse. The distance from the detector to the object can be determined based on the TOF theory.

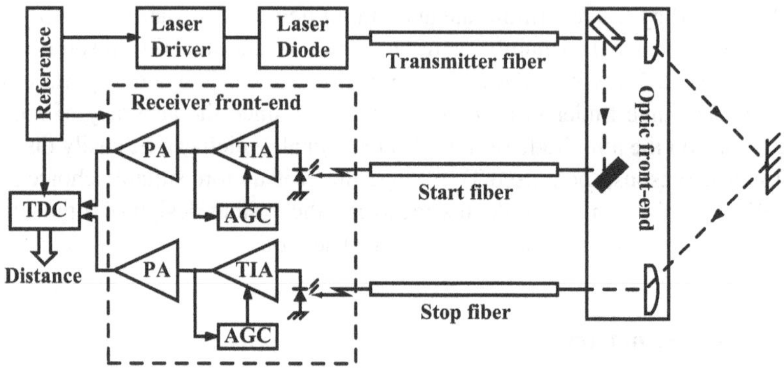

Fig. 1.5 Structure of a pulsed TOF laser range-finding system

1.4.2.1 Receiver Front End

The receiver front end consists of two identical channels, one for the start pulse and the other for the stop pulse. A photodetector converts the optical pulse to a current signal. As this current pulse is too weak to be processed directly, it is first amplified to a voltage signal by a low-noise transimpedance amplifier(TIA). It is then further amplified by a post amplifier to a level which can be detected by a time discriminator.

The simplest approach to detect the arriving time of the voltage pulse is the leading edge discrimination. It only uses a comparator to sense the moment when the output signal of the post amplifier exceeds a certain threshold [61]. The most significant advantage of using leading edge timing detection is that a very large dynamic range can be achieved. It is not restricted by the linear range of the receiver channel, as the timing information is contained only in the rising edge of the input pulse. However, the performance of the leading edge technique is often limited by the large walk error caused by the varying amplitude of the input signal. This is illustrated in Fig. 1.6. *vp1* and *vp2* represent two voltage pulses with different amplitude. The moment when *vp1* or *vp2* crosses a predefined threshold voltage Vth is obviously signal dependent.

The walk error can be reduced by applying a compensation curve, which shows the walk error as a function of the measured amplitude or slew rate the received signal. More details regarding the compensation scheme is explained in [58]. An alternative approach to mitigate the walk error instead of calibrate it is adopting an unipolar-to-bipolar pulse shaping technique [45]. The high-pass filtering pulse shaper converts the unipolar pulses to bipolar ones, in which the time position of the zero crossing point is independent of the amplitude of the incoming signal, as shown in Fig. 1.6. The main disadvantage of using this approach is a linear receiver channel is required to process the signal. This will limit the receiver's dynamic range under large echo signal circumstances.

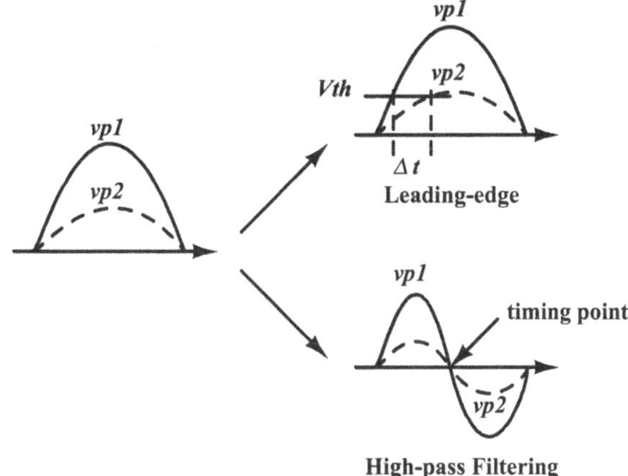

Fig. 1.6 Illustration of the walk error

1.4.2.2 Time-to-Digital Conversion

The accuracy of a TOF range finder is also limited by the resolution of the TDC employed. For instance, in order to detect 1 mm distance, a TDC with at least 6.7 ps time resolution is needed. Moreover, a TOF range finder for nuclear instrument applications typically works in a range of 1–30 m, which corresponds to a TDC measurement range of 3.3–200 ns. Therefore, a TDC with both wide input dynamic range and high resolution is required for this application.

The easiest way to perform time-to-digital conversion is using just a counter to count the pulses from a stable oscillator during the time interval to be determined. The resolution of such an arrangement is nevertheless limited by the period of the reference oscillator. Thus, to achieve a single-shot precision of 1 mm in distance measurement, a reference oscillator with a frequency of 150 GHz is needed. However, producing such a high frequency with small jitter in CMOS technology is quite challenging and power consuming. In the past decade, various TDC architectures have been investigated to achieve *ps*-level resolution. More details regarding the design of high-performance TDCs will be addressed in Chaps. 2 and 4.

1.5 Book Organization

Previous efforts have successfully implemented a laser driver [50] and a TIA [90] in CMOS technology with high radiation tolerance, which could partially form a radiation hardened integrated LIDAR system apart from the absence of a radiation-tolerant high-resolution TDC.

In this work, a megagray-radiation-tolerant *ps*-resolution TDC has been developed. For the first time, it successfully implements a third-order noise-shaping concept in the design of TDCs. The full TDC chip consists of three major parts

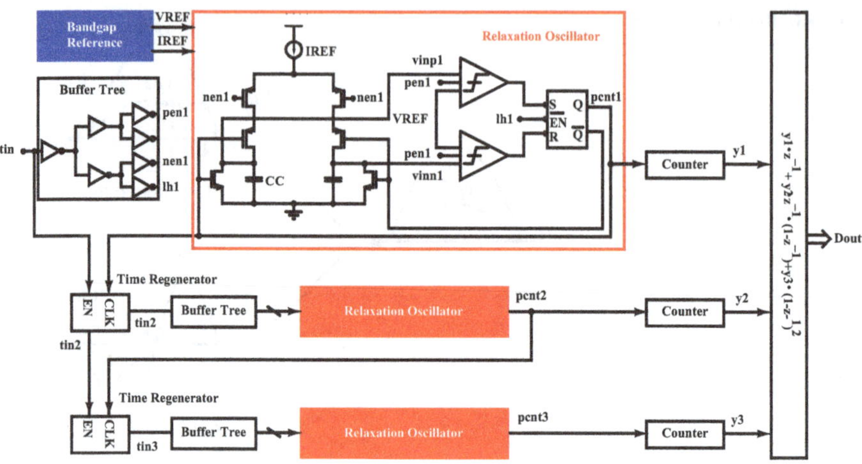

Fig. 1.7 System architecture of a radiation-tolerant MASH 1-1-1 $\Delta\Sigma$ TDC

(see Fig. 1.7): the three-stage noise-shaping signal processing structure, the relaxation oscillator (shown in red), and the radiation-hardened bandgap reference (shown in blue).

In Chap. 2, some background knowledge of TDCs is given. Chapter 3 introduces an overview of radiation effects on integrated circuits. It leads to a methodology study of radiation hardened by design. Chapter 4 explains the design process of a third-order noise-shaping TDC and an online calibration technique. The radiation tolerance of CMOS bandgap references is discussed in Chap. 5, where a radiation-hardened bandgap reference with dynamic base leakage compensation technique is proposed. Finally, the stability and accuracy issues of relaxation oscillators are investigated in Chap. 6.

Chapter 2
Background on Time-to-Digital Converters

Abstract Time-to-digital converters (TDCs) are key building blocks in time-based mixed-signal systems, used for the digitization of analog signals in time domain. Some commonly used TDC structures are summarized in this chapter. By carefully analyzing their advantages and performance constraints, it has been concluded that in order to design a megagray radiation-tolerant TDC, an innovative and reliable architecture must be developed. More technical details regarding the investigation of this new TDC type will be given in Chap. 4. In order to obtain a fare comparison between the TDC proposed in this book and other state-of-the-art TDCs, performance measures of TDCs are also given in this chapter.

2.1 Introduction

Recently, high resolution time-to-digital converters (TDCs) have gained more and more interest due to their increasing implementation in digital phase-locked loops (PLLs), analog-to-digital converters (ADCs), jitter measurement, and time-of-flight (TOF) range finders. A TDC is a device used to measure a time interval and convert it into digital output. In its simplest implementation, a TDC can be simply a high-frequency counter that increments every clock cycle. However, the time resolution of this type of TDC is practically limited by the highest available clock frequency.

In the next section, an overview of other state-of-the-art fine time measurement methods with much better accuracy but smaller measuring range is given. Performance measures of different TDCs are also discussed in this chapter.

2.2 TDC Topologies

Quantizing the time interval between a start signal and a stop signal, and representing it as a digital code, is the basic task of a TDC. The first TDCs were actually performing in two steps: time-to-voltage conversion (TVC), followed by voltage-to-digital conversion (VDC), as in [69, 86] (Fig. 2.1). The time signal is mapped into an analog voltage in the first phase, by using a charge pump. The amplitude of the voltage corresponds to the width of the time frame. In the second step, this voltage is translated into a digital code by a conventional ADC. A 30 ps resolution is the

© Springer International Publishing Switzerland 2015
Y. Cao et al., *Radiation-Tolerant Delta-Sigma Time-to-Digital Converters*,
Analog Circuits and Signal Processing, DOI 10.1007/978-3-319-11842-0_2

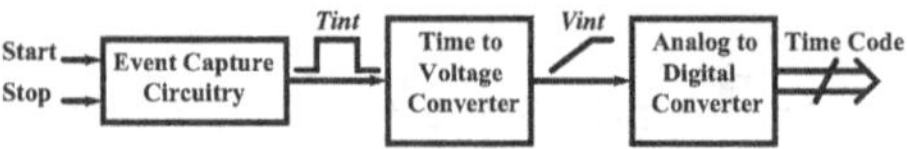

Fig. 2.1 Block diagram of the TVC–VDC architecture. *TVC* time-to-voltage conversion, *VDC* voltage-to-digital conversion

best that has ever been reported based on this configuration [69]. The performance is mainly limited by the nonlinearity in the TVC unit and the resolution of the ADC. Moreover, as technologies scale down, this approach becomes less attractive. The strongly scaled supply voltage but relatively unchanged threshold voltage, raises significant challenges in the design of a high-performance ADC, and restricts the input range of the TDC.

As opposed to the traditional analog method, a TDC could be also designed in time domain, where a circuit gains most profits from technology downscaling in terms of speed, power consumption, and area. The simplest time domain TDC is a counter. By using a high-speed low-jitter reference clock, the counter can digitize a time signal with moderate resolution. However, when the requirement for the TDC resolution is increased to a few picoseconds, the reference clock frequency becomes unreasonably high (>100 GHz) with respect to power consumption and system complexity. Therefore, a real TDC employs phase-aligned parallel counting clocks to achieve high resolution rather than using a single external reference clock. A complementary metal–oxide–semiconductor (CMOS) gate delay line can serve this purpose. Many TDC architectures based on a delay line core are reported in the past decade, and some achieve better than 10 ps resolution. They can be summarized into following categories:

- Flash TDC [25, 37, 38, 41, 82, 95].
- Pipeline TDC [49, 74].
- Successive approximation TDC [53].
- Noise-shaping TDC [83].

Similar to their ADC counterparts, each type of TDC performs well in one area, e.g., resolution, bandwidth, robustness, or power consumption, and lacks in another, as will be discussed below.

2.2.1 Flash TDC

A flash TDC uses a linear delay-cell ladder with a D flip-flop (DFF) at each rung of the ladder to compare the input time signal to successive reference time units [82], as shown in Fig. 2.2. A start signal propagates along the delay line, and the state of each delay element is sampled on the rising edge of the stop signal. A thermometer code is then generated at the DFFs' output, which represents the time difference between the start and stop signals. The timing and waveforms at the input of each DFF are illustrated in Fig. 2.3. The advantages of this circuit are obvious. It employs a very simple structure with only delay cells and DFFs. Hence, it is very area efficient. The

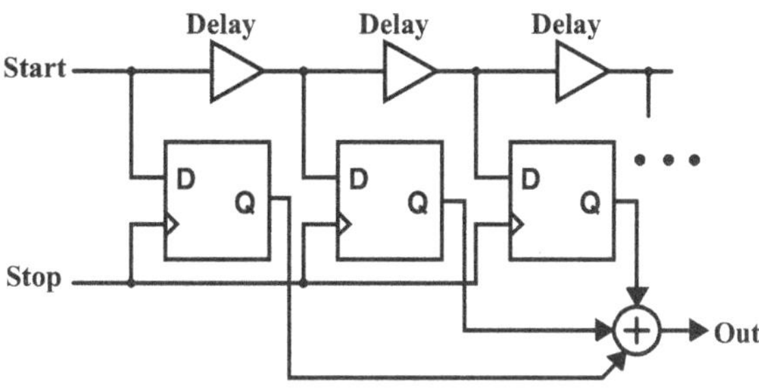

Fig. 2.2 Basic structure of a delay-line-based flash TDC. *TDC* time-to-digital converter

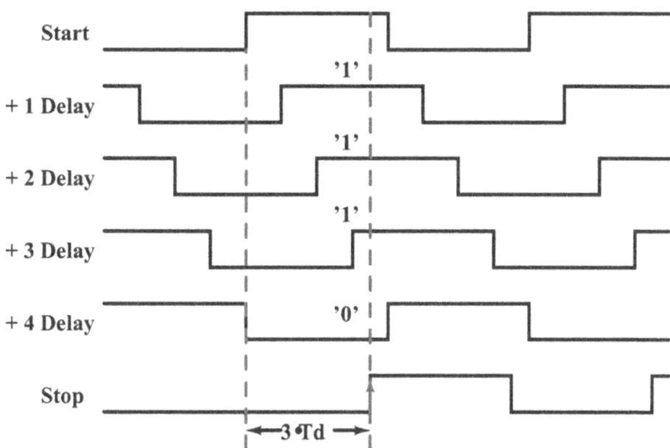

Fig. 2.3 Timing of the delay-line-based flash TDC. *TDC* time-to-digital converter

resolution is determined by the intrinsic CMOS gate delay, which scales according to the technology scaling factor. But practically, the TDC resolution does not continue to improve with technology downscaling, due to the worsened mismatch problem between delay cells. Therefore, the finest achievable resolution of the basic delay-line TDC is limited to around 20 ps.

In order to obtain sub-gate-delay resolution, the Vernier method [25] is commonly used, as shown in Fig. 2.4. Instead of using only one delay chain, the Vernier TDC utilizes two independent delay lines on both start and stop signal paths to improve the time resolution. The delay elements on the stop path are designed slightly faster than those on the start path. Both start and stop signals propagate along two delay lines with an initial time difference of T_{in}. The conversion is completed only after the stop signal outruns the start signal. The resolution of the Vernier TDC is then given by $t_{\text{delay1}} - t_{\text{delay2}}$. However, along with the resolution, the sensitivity of the Vernier TDC to mismatch has also been amplified. Although calibration can be applied to compensate for this error [37], significant efforts are needed since each delay element in the TDC has to be corrected individually.

Another technique to improve TDC resolution below that of a gate delay is to subdivide the coarse time interval given by an inverter delay line. This concept can be realized by placing a resistor divider between two nodes of an inverter, as presented in [38]. The divider interpolates the input and output signals of the digital gate, and creates new intermediate signals which effectively divide the gate delay into smaller intervals. The improvement in resolution for the interpolation architecture over the

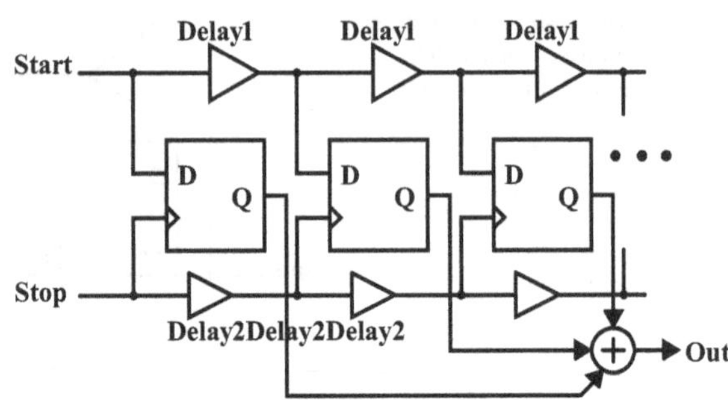

Fig. 2.4 Basic structure of a Vernier delay-line TDC. *TDC* time-to-digital converter

gate delay is similar to that of the Vernier architecture, and is practically limited by the nonlinear impedance of the delay elements during signal transients.

Since a flash TDC works at the Nyquist conversion rate, a large signal bandwidth can be obtained. High resolution is achievable at the cost of increased power consumption and area penalty. In practice, the random variations among delay elements set up an upper bound for the resolution of a flash TDC. To obtain well-controlled delay cells, the basic delay line core can be placed in a delay locked loop (DLL), as suggested in [41].

2.2.2 *Pipeline TDC*

It is well-known that, a pipeline ADC uses two or more steps of subranging and residue amplification technique to achieve high resolution. The same idea can also be realized in the design of TDCs, with the help of a time amplifier (TA). The working principle of the TA is illustrated in Fig. 2.5. An set–reset (SR) latch followed by an exclusive-OR (XOR) gate, is the main building block of the TA. If rising edges are applied to S and to R at almost the same time, the latch will be metastable. The initial voltage difference developed at the output of the SR latch is proportional to the input initial time difference ΔT_{SR}. The positive feedback in the latch forces the outputs of both NAND gates eventually to a binary level, which toggles the XOR gate to 1, and complete the regeneration process. The relationship of the regeneration time and the initial time difference is a logarithmic function as shown in Fig 2.5c.

Figure 2.6 shows the conceptual diagram of a pipeline TDC, which was proposed in [49]. First, the input time signal is digitized by a coarse flash TDC. The conversion result is then converted back to a reference time and subtracted from the original input. The residue time is then applied to another flash TDC after amplification. The effective resolution of the second flash TDC is thereby improved by a factor of the gain of the TA. However, unlike voltage, the residue time cannot be stored unless it is transformed to other forms such as voltage or current. Therefore, in a pipeline TDC, every possible time residue must be created and amplified separately. This significantly increases the system latency, and limits the input range of the TDC, since the linear working region of the TA is quite restricted. Moreover, the gain of the TA is also very sensitive to its working environment, mismatch, and process variation.

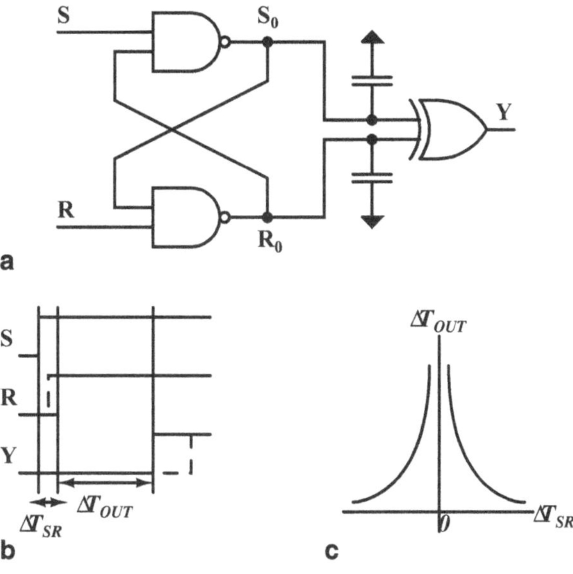

Fig. 2.5 **a** Basic architecture, **b** timing diagram, and **c** relationship between the regeneration time and the initial time difference of the time amplifier

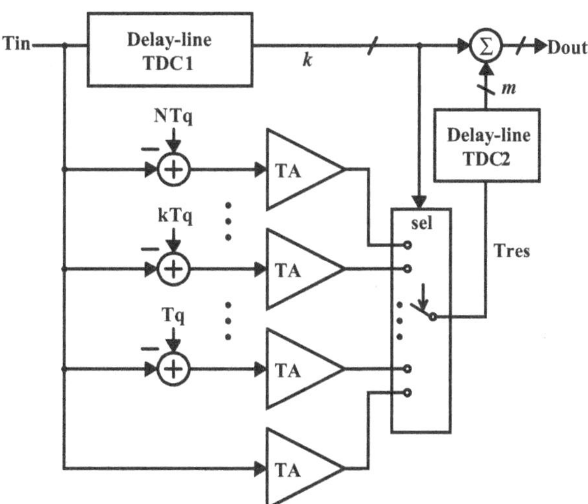

Fig. 2.6 Conceptual diagram of the TA-based pipeline TDC. *TA* time amplifier, *TDC* time-to-digital converter

2.2.3 Successive Approximation TDC

Successive approximation has been widely used in the design of ADCs to reach high resolution at the cost of conversion time. In the time domain, a successive approximation TDC [53] resolves the time difference between the start and stop signal one bit at a time in N cycles using binary search, as illustrated in Fig. 2.7. Due to the irretrievable nature of a time signal, the bidirectional adjustment required by the binary search is implemented by making both signal paths adjustable, rather than

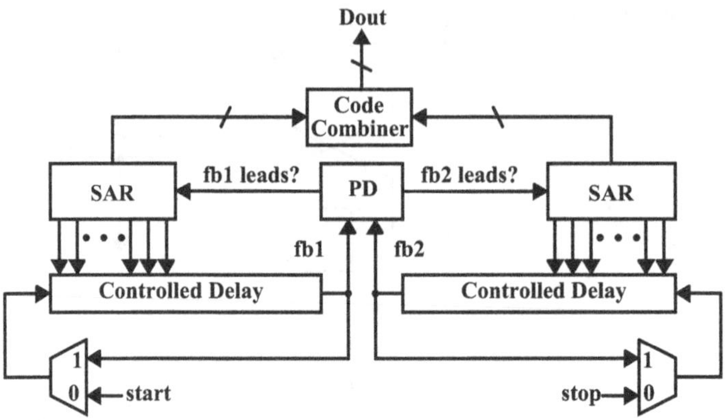

Fig. 2.7 Signal flowchart of the successive approximation TDC. *TDC* time-to-digital converter

adjusting only one signal back and forth. The two delayed versions of input signals, *fb1* and *fb2,* propagate cyclically in two separate loops formed by digital-to-time converters (DTCs), whose delays are controlled by the successive approximation register (SAR). At the beginning of the conversion, the DTC at the start path has a delay of $T_{\mathrm{REF}}/2$. The relative timing of *fb1* and *fb2* is compared with a phase detector (PD) to determine which signal is lead. The SAR will adjust its value according the output of the PD. Whenever the signals *fb1* and *fb2* are aligned within one LSB, the conversion is complete. The fine resolution of the SAR TDC is obtained by interpolation. For an 8-bit operation, 128 unit interpolators (e.g., capacitors) are needed in one DTC, which occupies large area. In order to achieve wide input range, the SAR TDC has to be configured as a coarse–fine architecture, which has more severe matching problems and consumes more power.

2.2.4 GRO TDC

A gated ring oscillator(GRO) TDC with first-order noise shaping has been reported in [83]. It can be considered as a first-order $\Delta\Sigma$ TDC. In the GRO TDC, the input time signal is used to enable/disable a ring oscillator, as illustrated in Fig. 2.8. One single measurement is done by counting all the phase transitions in the oscillator during the enabling phase. The quantization error, which refers to the intermediate state of the oscillator, is preserved between measurements. This results in a first-order noise shaping on the quantization noise. After digital low-pass filtering, the signal can be reconstructed with strongly reduced quantization error. The theory of time domain noise shaping will be discussed more in detail in Chap. 4. One issue in the GRO TDC, which could completely disrupt the noise-shaping behavior, is the existence of large skew error, as shown in Fig. 2.9. Caused by the charge redistribution during the "silent" phase of the TDC, the skew error results in imperfect preservation of the

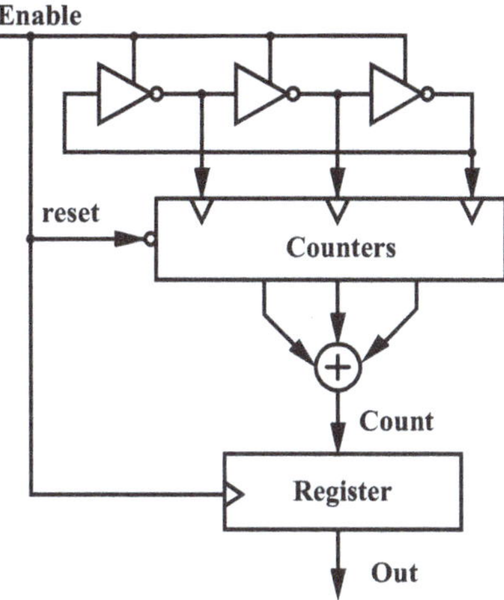

Fig. 2.8 Basic structure of the GRO TDC. *GRO* gated ring oscillator

quantization error. Another drawback of the GRO TDC, is the difficulty of achieving high-order noise shaping with this structure, which could principally further improve the TDC resolution and reduce the need for fast delay elements.

2.3 Performance Measures

2.3.1 Raw Resolution

The raw resolution of a TDC is the minimum quantization step of the system. In delay-line-based flash TDCs, this is equal to the delay time of one single delay cell, T_d. In counter-based TDCs, the raw resolution equals to one period of the counting clock, T_{OSC}.

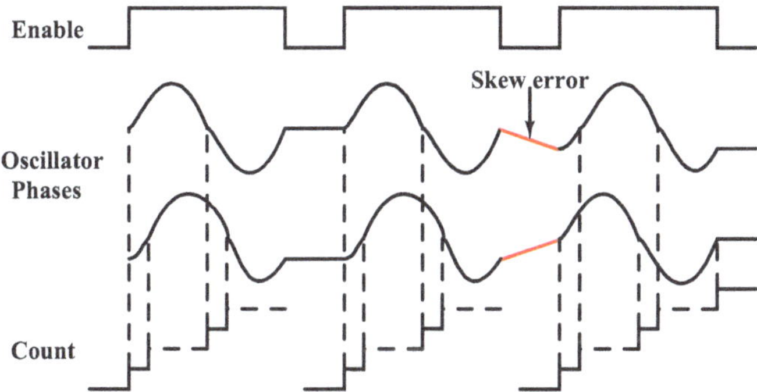

Fig. 2.9 Timing diagram of the GRO TDC. *GRO* gated ring oscillator

2.3.2 Single-Shot Precision

If a constant time interval is measured repeatedly, the digital output values vary with a standard deviation that is called single-shot precision. In certain circumstances, a very small single-shot precision is of importance when repeated measurement is not possible. For instance, in the CMS experiment mentioned in Chap. 1, the time of-arrival of generated particles after a collision event needs to be measured. Apparently, there is only one chance to conduct an accurate measurement before the particle has fleeted out.

However, for many other industrial applications, repeated measurement of the same time interval is feasible as soon as the repeating frequency is much higher than the varying frequency of the input time signal's amplitude. In the LIDAR experiment, this is mostly the case. Therefore, the single-shot precision is not focused during the design of the radiation-tolerant $\Delta\Sigma$ TDC for LIDAR application.

2.3.3 Sampling Rate

In order to achieve a stable measurement of the input coming signal, a TDC is often configured in a closed-loop system. For instance, in a basic flash TDC, the delay line is placed in a DLL, which reduces nonlinear errors caused by mismatch. In such a closed-loop structure, the system is refreshed at a fixed frequency, which is defined as the sampling rate (F_s) of the TDC.

2.3.4 Bandwidth

The bandwidth of a TDC is the range of input signal frequencies it can measure. For flash TDCs, this is mostly half of their sampling frequency according to Nyquist theorem. For oversampling TDCs, it is equal to $F_s/(2{\cdot}\mathrm{OSR})$, where OSR is the oversampling ratio.

2.3.5 Effective Resolution

The effective resolution is the actual resolution of a TDC when circuit noise and mismatch effects are taken into account. In flash TDCs, the effective resolution is usually worse than its raw resolution. However, in $\Delta\Sigma$ TDCs, the raw resolution can be further improved by oversampling and noise shaping. Therefore, within the interested signal bandwidth, the noise-free effective resolution of a first-order $\Delta\Sigma$

TDC can be expressed as

$$T_{\text{eff}} = \frac{\pi \cdot T_{\text{d}}}{6 \cdot \sqrt{\text{OSR}^3}}, \tag{2.1}$$

where OSR is the oversampling ratio, and T_{d} is the raw resolution. Of course, when circuit noises are added into the figure, the real effective resolution will be worse.

2.3.6 SNDR and ENOB

SNDR is defined as the signal to noise plus distortion ratio within the interested signal bandwidth. It is the major figure which is used to measure a TDC's dynamic performance:

$$\text{SNDR} = \frac{P_{\text{signal}}}{P_{\text{quantizationerror}} + P_{\text{radomnoise}} + P_{\text{distortion}}}, \tag{2.2}$$

where P is the average power of the signal, quantization error, circuit random noise, and distortion components.

Effective number of bits (ENOB) specifies the number of bits in the digitized signal above the noise floor. SNDR (in dB) is related to ENOB by the following equation:

$$\text{SNDR} = \text{ENOB} \cdot 6.02 + 1.76. \tag{2.3}$$

2.3.7 Dynamic Range

Similar to an ADC, the dynamic range (DR) of a TDC is defined as the maximum time interval that can be measured without any saturation effects. In the LIDAR experiment, a large DR is more preferred since the system's detection range needs to go over 30 m, meanwhile, the required resolution is 1 mm. It turns to a DR of

$$\text{DR} = 20 \cdot \log\left(\frac{A_{\text{fullrange}}}{A_{\text{noise}_{\text{rms}}}}\right) = 20 \cdot \log\left(\frac{30}{1 \times 10^{-3}}\right) = 90\ \text{dB!} \tag{2.4}$$

Chapter 3
Radiation Hardened by Design

Abstract The major work introduced in this book is to develop a megagray-radiation-tolerant *ps*-resolution time-to-digital converter (TDC) for a light detection and ranging (LIDAR) application. In this chapter, major radiation effects in complementary metal–oxide–semiconductor (CMOS) integrated circuits (ICs) are introduced. The total ionizing dose (TID) effects are of most concern to this work since they have the largest impact on analog performance of a system. Hierarchical radiation hardened by design (RHBD) strategies are then proposed to improve the radiation tolerance of CMOS devices from four levels: system level, circuit level, device level, and layout level. Finally, the radiation hardness assurance qualification procedure is also given, which has been followed during TID performance evaluation of the demonstrated CMOS ICs presented in this book.

3.1 Introduction

In the past, radiation hardening of electronic components is achieved via process modification by rad-hard foundries, which is so-called radiation hardened by process (RHBP). While RHBP has the advantage of being an extremely reliable means of achieving hardened components, it is susceptible to low-volume concerns such as yield, process instability, and high manufacturing costs [46].

In order to leverage these limitations, the RHBD approach was proposed. In RHBD, electronic components are manufactured to meet specified radiation tolerance requirement, but the techniques employed to meet these specs are implemented either in the system architecture or in layout and not in the fabrication process. Through a combination of the application of specific design techniques and the leveraging of the increased intrinsic radiation hardness of modern advanced integrated circuit (IC) technologies, it is now possible to fabricate radiation-hardened components using standard complementary metal–oxide–semiconductor (CMOS) processes.

In this chapter, a systematic elaboration of several most useful RHBD techniques will be addressed. First, an overview of radiation effects on CMOS ICs is given in Sect. 3.2. Section 3.3 discusses specific approaches to improve radiation tolerance of ICs from system level to layout level. A brief introduction of the radiation hardness assurance qualification procedure is also provided in Sect. 3.4.

© Springer International Publishing Switzerland 2015
Y. Cao et al., *Radiation-Tolerant Delta-Sigma Time-to-Digital Converters*,
Analog Circuits and Signal Processing, DOI 10.1007/978-3-319-11842-0_3

3.2 Radiation Effects in CMOS ICs

Radiation effects in CMOS ICs can be divided into three major categories: total ionizing dose (TID) effects, single event effects (SEEs), and displacement damage . TID effects and SEEs both are from ionizing radiation, while displacement damage is the result of nuclear interactions, typically scattering, which causes crystal lattice defects. The major consequence of displacement damage in a semiconductor material is increasing of the number of recombination centers and depleting of the minority carriers. This type of problem is particularly significant in bipolar transistors, which are dependent on minority carriers in their base regions. It is typically of less concern than TID or SEEs for CMOS devices, which conduct current almost entirely by majority carriers. On the other hand, TID effects are mainly associated with interactions between the irradiated material and charged particles such as gamma rays, while SEEs are resulting from the interaction of highly energetic particles, e.g., protons or other heavy particles [47]. Thus, TID can be considered as a long-term failure mechanism versus SEE which is an instantaneous failure mechanism. In this section, we first introduce the TID effects in CMOS devices, and the SEE will be discussed in the last part of this section.

3.2.1 TID Effects in MOS Devices

High-energy photons or charged particles (i.e., protons, electrons, or energetic heavy ions) can ionize a material, generating electron–hole pairs. It is caused by the interaction of the high-energy particle with the atoms of that material [9]. When a MOS transistor is exposed to high-energy ionizing irradiation, electron–hole pairs are created in the oxide (both gate oxide and field oxide). Almost all TID effects in CMOS transistors are related to electron–hole pair generation in the oxide [71]. The generated carriers induce the buildup of charge, which can lead to device degradation. Figure 3.1 illustrates the cross section of the physical layout of a standard MOS transistor. The effects of TID radiation in an MOS device will be mainly on: (1) gate oxide, which can result in threshold voltage shift; (2) channel edges, which can result in turning on of the parasitic edge transistor; (3) isolation oxide, which can result in increased interdevice leakage, or even complete loss of device isolation.

Figure 3.2 shows a schematic energy band diagram of a MOS structure in a p-substrate, where positive bias is applied to the gate [60]. The four major physical processes are indicated, which contribute to the ionizing radiation response of a MOS device [71]. The most sensitive part of an MOS transistor to radiation is the gate-oxide insulator. When radiation passes through it, electron–hole pairs are created. Since in SiO_2, electrons are much more mobile than holes, they rapidly drift (within picoseconds) toward the positively biased gate immediately after the creation. However, even before the electrons are swept out of the oxide, some of the electrons will recombine with holes. The holes, which escape initial recombination, will move

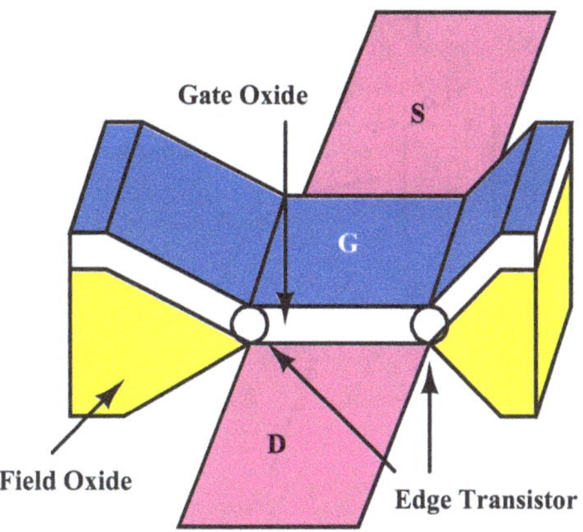

Fig. 3.1 Illustration of the physical layout of a standard MOS transistor

through the oxide toward the Si/SiO_2 interface by hopping through localized states in the oxide. As the holes approach the interface, some fraction will be trapped, forming a positive oxide-trap charge. The trapped holes cause a negative threshold voltage shift, which can persist for hours or even for years. Taking into account the hole yield and initial electron–hole recombination, the total number of holes generated in the oxide N_{ot} is given by [55]:

$$N_{ot} = f(E_{ox}) \cdot g_0 \cdot D \cdot t_{ox}, \tag{3.1}$$

where $f(E_{ox})$ is the hole yield as a function of oxide electric field, D is the total dose, and t_{ox} is the oxide thickness. g_0 is a material-dependent parameter giving the initial charge pair density per rad of a dose ($g_0 = 8.1 \times 10^{12}$ pairs/cm^3 per rad for SiO_2 [55]).

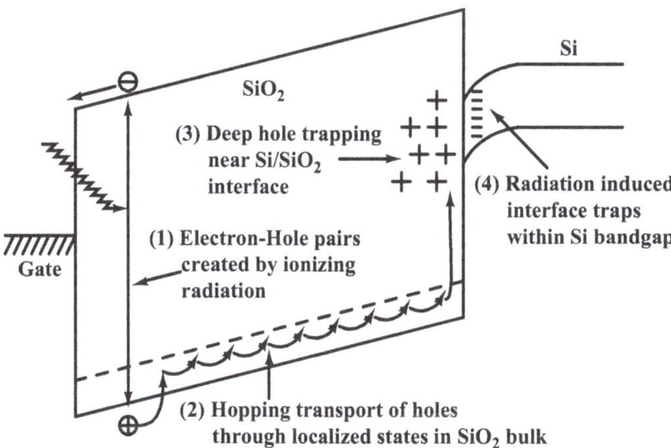

Fig. 3.2 Illustration of the energy band diagram of an MOS structure with a positive gate bias, and major physical processes of TID irradiation

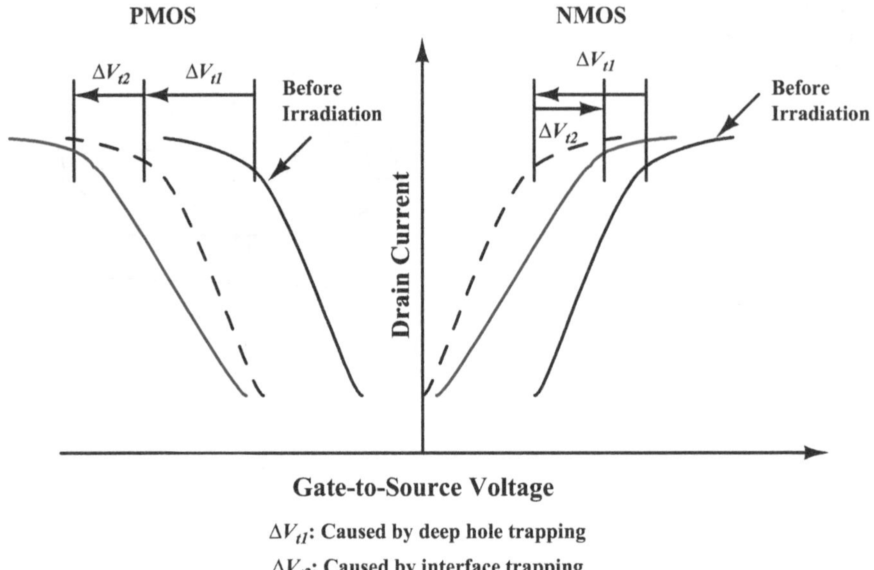

Fig. 3.3 Illustration of the effect of trapped charges near the Si/SiO_2 interface on n-type metal-oxide-semiconductor (*NMOS*) and p-type metal-oxide-semiconductor (*PMOS*) transistors.

The last major process of the MOS radiation response is the radiation-induced buildup of interface traps right at the Si/SiO_2 interface. Interface traps exist within the silicon band gap at the interface, and they can be positive, neutral, or negative [71]. Traps in the lower portion of the band gap are predominantly donors, i.e., if the Fermi level at the interface is below the trap energy level, the trap "donates" an electron to the silicon. In this case, the trap is positively charged. p-Channel transistors have a low Fermi level in the surface inversion layer when at threshold, therefore they are affected primarily by interface traps in the lower region of the band gap. Conversely, n-channel transistors are affected primarily by interface traps in the upper region of the band gap. Interface-trap buildup occurs on time frames much slower than oxide-trap charge buildup, and it can take thousands of seconds to saturate after a pulse of ionizing radiation [76].

Those trapped charges near the Si/SiO_2 interface can have a significant impact on the direct current (dc) parameters of CMOS devices. One of the most important effects is the threshold voltage shift for both n-channel and p-channel metal-oxide-semiconductor field-effect transistor (MOSFETS) [9]. This effect is illustrated in Fig. 3.3. In NMOS transistors, the voltage shift caused by the oxide-trapped holes leads to a reduction in threshold voltage and an increase in off-state current. In PMOS transistors, the threshold voltage increases negatively, while off-state current is reduced. For the interface traps, since they are predominantly positively charged for PMOS transistors and negatively charged for NMOS transistors, they will cause negative threshold voltage shift for PMOSFET and positive threshold voltage shift

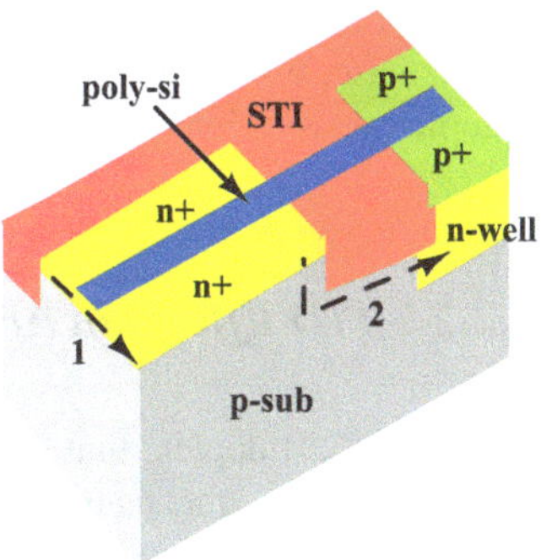

Fig. 3.4 Indication of *1* the parasitic edge transistor leakage path and *2* the interdevice leakage path in CMOS technology

for NMOSFET, respectively. Thus, the total threshold voltage shift ΔV_{th} is the sum of the threshold voltage drifts due to oxide-trapped charge and interface-trapped charge, given by

$$\Delta V_{th} = \Delta V_{ot} + \Delta V_{it}. \tag{3.2}$$

Meanwhile,

$$\Delta V_{ot} \propto \frac{N_{ot}}{C_{ox}} \propto t_{ox}^2. \tag{3.3}$$

where N_{ot} is the total oxide-trapped charge and is proportional to t_{ox} as described in (3.1), and C_{ox} is inversely proportional to t_{ox}.

At high dose rates, neutralization of oxide-trapped charge will hardly occur and ΔV_{ox} can be large. Conversely, interface-trapped charge will have had insufficient time to build up and ΔV_{it} is normally small [71]. Therefore, the total threshold-voltage shift is usually dominated by ΔV_{ot} in this case.

Another important TID effect is the radiation-induced edge effects. As illustrated in Fig. 3.1, transistors in conventional CMOS technologies are isolated by thick field oxide. The combination of the extended gate poly, the isolation field oxide, and the diffusion region form a parasitic edge transistor. It also exists in modern CMOS technologies where transistors are isolated by shallow-trench isolation (STI). When a NMOS transistor is exposed to radiation, the positive trapped charges in the isolation oxide will result in a negative shift of the threshold voltage. This will gradually turn on the edge transistor, which contributes a significant leakage current to the main transistor (as indicated in Fig. 3.4). However, edge leakage is not a problem for PMOS transistors, since for PMOS transistors the effect of oxide-trapped positive charge is to shift an already negative threshold voltage even further in the negative direction.

The buildup of positive charges in the field oxide also poses problems on interdevice isolation. The isolation oxide and two closely placed n-type diffusion regions from different devices or wells also potentially form a parasitic NMOS transistor. If a conducting line (metal or poly) happens to be routed over the isolation oxide area, any positive bias presented on the line might turn on this parasitic field oxide transistor into inversion due to the radiation-induced negative threshold voltage shift. In this case, a leakage path exists between different devices, which might lead to complete loss of interdevice isolation.

3.2.2 *TID Effects in Advanced CMOS Technologies*

Equation (3.3) depicts that the radiation-induced threshold voltage shift is proportional to t_{ox}^2. This means that CMOS technology scaling would reduce a MOSFET's susceptibility to the radiation-induced damage in gate oxide [9]. This is primarily due to the fact that the trapped charge buildup in gate oxides scales with t_{ox} (3.1). Moreover, the positive trapped holes are further annihilated or compensated by tunneling electrons from the gate channel [20]. Enhanced CMOS gate oxide hardness (on the order of 1–3 kGy(SiO_2)) has been observed at small technology nodes, e.g., 0.18 and 0.13 μm CMOS [46]. At these nodes, gate oxide thicknesses t_{ox} are smaller than 4 and 2 nm, respectively, which are very close to the approximate distance for a high-probability electron tunneling ($\sim$3 nm) [20]. The increased TID hardness of commercial CMOS technologies has been proved experimentally and reported in several publications [48, 66].

The above analysis suggests that as technology scaling continues, the radiation-induced threshold shift of CMOS transistors would become negligible. This is also the primary reason that RHBD has become a realistic option over the past few years, while more circuits were designed at smaller technology node (<0.18 μm). As soon as transistors under irradiation can maintain being functional without significant threshold voltage shifts, all other radiation effects (e.g., leakage, gain error, offset, etc.) can somehow be mitigated through compensation or calibration techniques.

While the gate-oxide-trapped charge appears to be no longer an issue for advanced CMOS technologies, radiation-induced charge trapping in the isolation field oxide still leads to increased leakage currents and device performance degradation [27, 42]. This is mainly due to the fact that the STI oxide of modern CMOS technologies does not scale down with the gate size to the same extent. Nevertheless, it is possible with RHBD circuit or layout techniques demonstrated in Sect. 3.3 to eliminate these effects.

3.2.3 *Single Event Effects*

Single event effects in CMOS devices are also caused by ionizing radiation. Unlike the TID effects, SEEs are mainly due to the interaction of the semiconductor material with heavy ions. Lighter particles, such as protons and neutrons, usually do not produce enough charge to cause SEE in a CMOS device. However, these particles may undergo an inelastic collision with a target nucleus in the semiconductor to produce heavier particles through spallation reaction, which can induce SEEs [47]. When an energetic heavy ion hits a semiconductor device, it will create highly concentrated electron–hole pairs along its path in the device's body. If the ionization track traverses the depletion region of a reverse biased p–n junction (e.g., the junction between a n-type diffusion area and the p-type substrate), carriers will rapidly drift by the electric field present in the depletion region and be collected at the respective contact nodes. This will induce an effective short across the p–n junction for the period of time needed to dissipate the generated charge, and create a large current/voltage transient at that node [11].

In general, digital circuits are more susceptible to SEEs, since their functionalities are premised on correct storage of intermediate register status, which might be disturbed by large current/voltage transients caused by SEEs. Commonly seen single event effects in a CMOS circuit can be categorized into three types: single event upsets (SEUs), single event transients (SETs), and single event latchup (SEL). An SEU is the consequence of an energetic particle strike directly on a critical node which causes data to change states in a storage element such as a flip-flop, latch, or memory bit. An SET addresses errors that can result from an energetic particle strike on non-latched elements, such as combinational logic. SEL occurs when a particle strike triggers the parasitic semiconductor-controlled rectifier (SCR) structure in a CMOS circuit.

Since the work presented in this book includes mainly analog circuitries and is intended to be used in an environment with predominant high-dose-rate gamma rays, the TID effect is of primary concern. SEEs will not be discussed in this book, even though precautions have been taken during the design of digital blocks employed in the time-to-digital converter (TDC) system. For example, intensive guard rings were used in the layout of digital cells to ensure their immunity to SEL.

3.3 Radiation Hardened by Design

Over the course of the past decade, the TID radiation hardness of commercial CMOS technologies has been evolving rapidly, due to the continual shrinking of the gate oxide thickness. However, there is no guarantee of the same radiation tolerance level among different processes at the same technology node provided by different manufacturers, since the radiation hardness is not a parameter that commercial semiconductor foundries monitor [22]. Moreover, the TID effects in a given CMOS technology are very much linked to the exact thickness of the gate oxide, imperfection levels in the isolation SiO_2 region, etc. Unfortunately, those parameters are usually not modeled in CMOS devices, which makes the prediction of the impact of irradiation on a CMOS circuit very difficult. They are also often not disclosed by the manufacturer, complicating the semiconductor level modeling of radiation effects. Therefore, besides choosing a more advanced CMOS technology for radiation-hardened designs, RHBD techniques are also required in the circuit to ensure its reliability and performance even under an extreme radiation level.

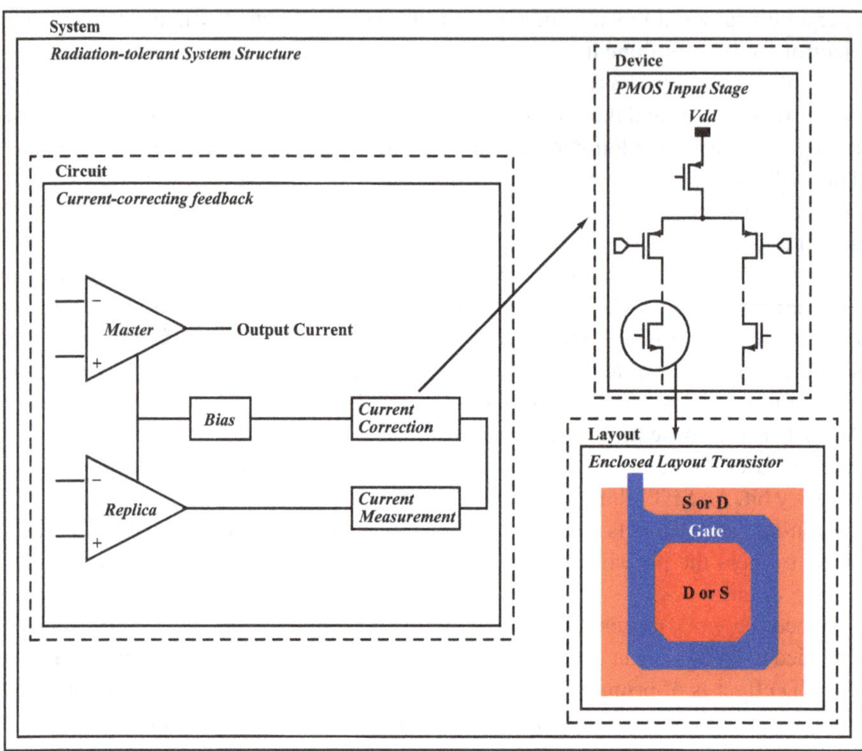

Fig. 3.5 Radiation hardened by design strategy

RHBD techniques can be employed during the whole design phase of a CMOS chip. In general, they can be separated into four hierarchical levels: system level, circuit level, device level, and layout level (as illustrated in Fig. 3.5). Each of them will be discussed in detail in the remainder of this section.

3.3.1 System-Level Approach

System-level RHBD is the most effective way to assure the radiation hardness of an electronic device. When an electronic system is required to perform certain functions in a radiation environment, one should first consider choosing a system structure which provides the highest radiation tolerance.

A good example is the selection of a suitable analog-to-digital converter (ADC) for radiation-hardened applications. ADCs have various architectures, such as the flash, pipeline, successive approximation (SAR), dual slope integrating, and delta–sigma type. For a given specification in terms of resolution, power consumption, and dynamic range, all those ADC types might be suitable. However, they indeed have different radiation tolerance levels. Several commercial ADC parts from different manufacturers were compared in [4] to examine their gamma radiation performance. Interestingly, a tested delta–sigma ADC (ADS1210) exhibits the highest radiation tolerance level (up to 25 kGy) when an external reference voltage is used for it (if the internal bandgap circuit is used for providing the reference voltage, it will fail much sooner only after 200 Gy). And the flash ADCs are most vulnerable to TID effects. They suffer from functional failure at relatively low accumulated dose.

In this work, a high-resolution TDC is required to achieve MGy radiation tolerance level, which poses significant challenges on selecting a suitable system architecture. After investigating various TDC structures described in Chap. 2, we decided to go for an innovative multistage noise-shaping TDC, which is considered a more reliable solution. More technical details regarding the design of this TDC will be given in Chap. 4.

3.3.2 Circuit-Level Approach

Despite an appropriate system architecture could offer better radiation tolerance, it might suffer from circuit-level failures due to radiation, such as the bandgap circuit failure in the test delta–sigma ADC case stated above. Therefore, one must identify all radiation susceptible components in the system, and make sure that they can achieve the required radiation tolerance level. For this reason, we need to apply RHBD techniques at the circuit level.

As an example, [50] presents a circuit-level RHBD approach to improve the radiation hardness of a CMOS VCSEL driver. The driver is implemented in 0.7 μm CMOS technology. The radiation induced threshold voltage shifts of the MOS transistors in the driver, would cause a drift in the VCSEL current. This is tackled with a current correcting feedback mechanism, as shown in Fig. 3.6. The goal of this feedback circuitry is to keep the forward current through the VCSEL constant. An extra replica of the driver is employed to reproduce the same current. Since the replica driver and the main driver are identical devices, any performance changes in the main driver due to TID effects will also be reflected in the replica device. A control loop continuously monitors this duplicate current and adjusts the bias of the driver in order to keep its output current constant. Combining some layout RHBD techniques, the VCSEL driver achieves a TID radiation tolerance level of 3.5 MGy.

In this book, circuit-level RHBD techniques are demonstrated in a bandgap voltage reference, which will be further discussed in Chap. 5. The reference voltage generated by the radiation-hardened bandgap reference has shown a variation of only 3 % after 4.5 MGy, while that of the conventional bandgap reference increases more than 15 % during irradiation.

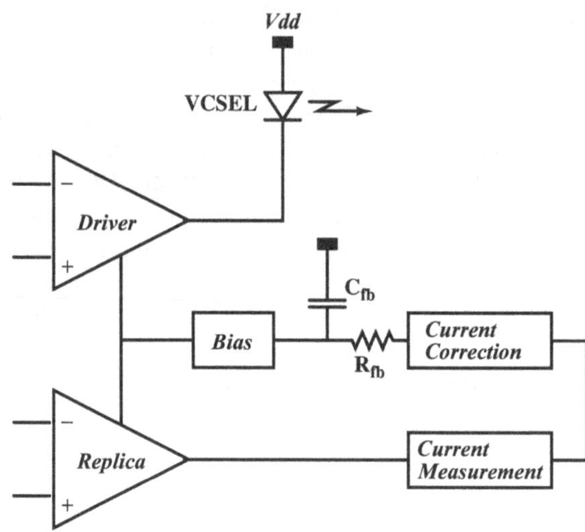

Fig. 3.6 Block level schematic of the radiation-hardened vertical-cavity surface-emitting laser (*VCSEL*) driver

3.3.3 Device-Level Approach

The device-level RHBD approach tackles CMOS radiation hardness assurance issues by selecting appropriate transistor types and geometries (width and length) for circuit implementation. For instance, in old CMOS technologies, PMOS transistors are preferred for radiation-hardened designs since their threshold voltage shifts only cause performance variations rather than functionality failures.

Moreover, a transistor's TID effects have also dependencies on its geometry. As described in [77] and [66], for the same technology, a transistor with longer gate-length shows much less radiation-induced threshold voltage shifts and leakage currents compared to short-gate-length devices. Therefore, during the actual circuit design phase, we have to take into account all these practical issues and carefully choose devices' types and geometries, e.g., using transistors with the minimum gate-length should be avoided. This will in return give you significantly enhanced circuit's radiation tolerance.

3.3.4 Layout-Level Approach

Layout-level RHBD techniques have been proved to be very effective in eliminating single-event latchup and preventing radiation-induced leakage currents. One example is the enclosed layout transistor (ELT) for ultra-high radiation-hardness applications [6]. An ELT has a closed gate shape, which separates the drain and the source region of the CMOS transistor. A typical layout of the ELT is shown in Fig. 3.7. Such a layout arrangement eliminates the parasitic transistors originally located at two edges of a conventional CMOS device. The radiation-induced MOS transistor off-state leakage

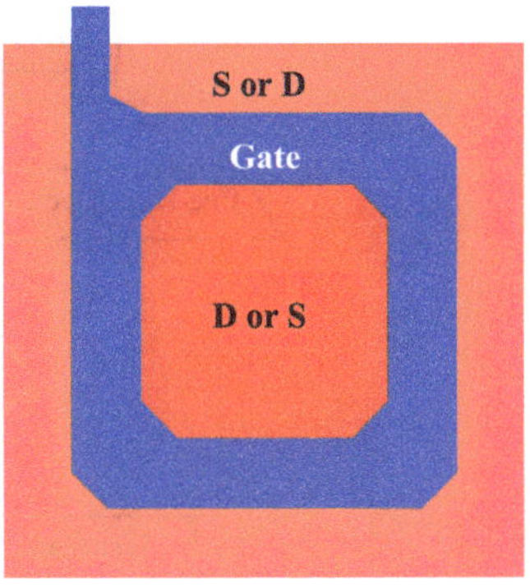

Fig. 3.7 Layout of the enclosed layout transistor

current has then been greatly reduced. Other homologous transformations of the standard ELT transistor layout can be found in [50, 81]. Main drawbacks of the ELT transistor are its relatively large area and inaccurate modeling.

In this work, ELT transistors are not used since all designs presented in this book were implemented in 0.13 μm CMOS technology or below, where edge transistor leakage currents are almost negligible. However, they are still useful devices in some applications where there is a zero tolerance to any radiation-induced leakage current, for instance, an NMOS switch connected to a charge storage element.

3.4 Radiation Hardness Assurance Qualification

There are a number of qualification test methods that define total dose testing of microelectronics. These include military standards (MIL-STD)-883, Test Method (1019) used in the USA and its European counterpart, European Space Agency/Space Components Coordination (ESA/SCC) Basic Specification (BS) No. 22900 [78]. Although both test methods are intended to provide qualification standard for semiconductor devices suitable for low-dose-rate space applications, they are also very instructive to the development of total dose irradiation procedure for microelectronics used in high-dose-rate applications.

The European Space Components Coordination (ESCC) 22900 total dose steady-state irradiation qualification procedure can be split into two phases, which are [26]:

- Evaluation of technology, especially oxide process variations and time-dependent effects.
- Qualification and lot acceptance of high reliability devices.

The main objective of the evaluation phase is to establish worst case conditions for TID qualification of a specific device/technology, and the lot acceptance qualification phase is to verify the absolute value and statistical spread of radiation tolerance in a device production line. A flow chart for the ESCC 22900 evaluation testing is shown in Fig. 3.8.

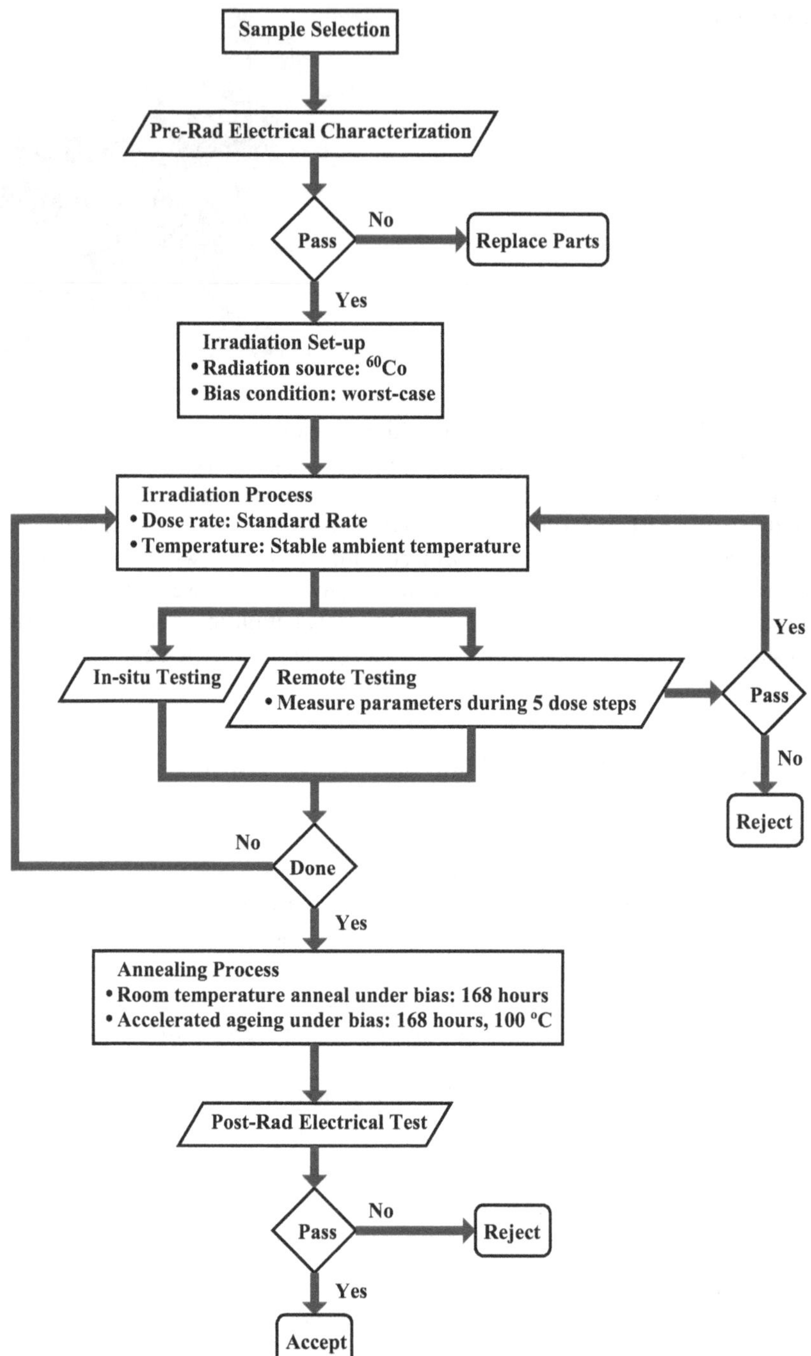

Fig. 3.8 Flow chart for the ESCC 22900 evaluation testing

In the first step, device samples prepared for the irradiation process need to be picked up. The main concern is that the radiation hardness of CMOS devices is technology dependent and has to be evaluated across process variations. According to ESCC 22900, a minimum sample of 11 test devices shall be selected at random from a minimum of two different diffusion lots, making a minimum of 22 samples in all. In-laboratory electrical characterization is then performed on all selected samples before irradiation.

During the irradiation setup, the radiation source for hardness assurance testing, radiation dose rates and bias conditions for the test devices are to be decided. The cobalt 60 (^{60}Co) gamma source can offer a wide range of radiation dose rates and is among the most commonly used laboratory radiation sources for radiation hardness assurance routine evaluation. The choice of the radiation dose rate is mainly based on two considerations, which are:

- The dose rate is preferred to be kept as close to the value in the real working environment. However, depending on the expected maximum total dose level, the dose rate shall be adjusted so that it can be achieved in a reasonable time scale.
- Principally, the dose rate shall be held constant during a given radiation exposure. However, in order to reveal any dose-rate-radiation effects, test devices should be evaluated under different dose rate settings (e.g., low rate (<3.6 Gy/h), standard rate (36–360 Gy/h), and high rate (>3.6 kGy/h)).

It is well known that the TID effects of CMOS ICs are largely dependent on the bias conditions. Thus, the bias applied to the test devices should be worst-case conditions to produce the greatest radiation-induced damage to those devices. The worst-case bias for advanced bulk CMOS technologies is the bias condition that maximizes the electric field across the field oxide, which is usually the supply voltage of the device [78].

Regarding the measurement of electrical parameters of test devices during irradiation, there are two different methods, which are directly measuring in the irradiation chamber (in situ testing) or after the removal from the chamber (remote testing). In situ testing is a better evaluation strategy since all intermediate radiation data can be recorded and no temporary effects will be missed out, such as the "radiation rebound effect." But in situ testing has significantly increased complexity and certain limitations compared to remote testing. For instance, since the test devices are mounted on a test circuit board together with other supporting components, all of them should be radiation tolerant in order to evaluate the test devices' radiation performance accurately. Moreover, as the irradiation chamber is normally located in a shielded area where it is further away from the measurement room, there must be long cable connections between measurement equipments and the test devices. This will pose serious noise coupling issues and dramatically deteriorate high-frequency signals. If remote testing is adopted, electrical parameters should be measured during multiple dose steps (at least five steps according to ESCC 22900). The time interval from the completion of an exposure to the start of the measurement of parameters should be a maximum of 1 h in order to avoid annealing of radiation effects in test devices [26].

After completion of the irradiation process, electrical parameters should be re-measured during room temperature annealing and accelerated ageing of test devices. A test device can be considered as radiation tolerant to a specific level only if its postirradiation electrical parameters have still met specifications.

Chapter 4
Background on Time-to-Digital Converters

Abstract For the first time, a third-order noise shaping concept has been successfully implemented in the design of time-to-digital converters (TDCs). Two 1-1-1 multi-stage noise shaping (MASH) $\Delta\Sigma$ TDCs are presented in this chapter. Third-order time domain noise shaping has been adopted by the TDCs to achieve better than 6 ps resolution. Following a detailed analysis of the noise generation and propagation in the MASH $\Delta\Sigma$ structure, the first prototyping TDC has been realized in 0.13 μm complementary metal–oxide–semiconductor (CMOS) technology. It achieves an effective number of bits (ENOB) of 11 bits and consumes 1.7 mW from a 1.2-V supply. Gamma radiation assessments with both a low dose rate of 1.2 kGy/h and a high dose rate of 30 kGy/h have been performed, proving the TDC's radiation hardness. In the second MASH TDC, a delay-line-assisted calibration technique is introduced to mitigate the phase skew caused by the large comparator delay, which is the main limiting factor of the MASH TDC's resolution. The demonstrated TDC achieves an ENOB of 13 bits and a wide input range of 100 ns.

4.1 Introduction

As a mainstream integrated circuit fabricating process, complementary metal–oxide–semiconductor (CMOS) technology has been successfully implemented under ionizing radiation up to 1 MGy, by laying out the N-type metal–oxide–semiconductor (NMOS) transistors in enclosed geometry [80]. Recent research also shows a trend in advanced CMOS technologies toward increased total dose hardness, due to down-scaling of the CMOS gate oxide thickness [47]. This makes modern deep-submicron CMOS technology more suitable for a radiation tolerant design. However, for upcoming applications in nuclear fusion reactors like the International Thermonuclear Experimental Reactor (ITER) [33], electronic components are required to achieve good performance in harsh environments facing high temperature and radiation, where the threshold voltage, transconductance, and delay of a transistor undergo dramatic changes [18]. In these cases, the high resolution, accuracy, and robustness of the time-to-digital converter (TDC) need to be inherent to the design.

As discussed in Chap. 2, conventional TDCs were built based on the CMOS gate-delay-line structure [82], whose highest achievable resolution is limited by the intrinsic delay of a CMOS inverter gate. In order to get sub-gate-delay

© Springer International Publishing Switzerland 2015
Y. Cao et al., *Radiation-Tolerant Delta-Sigma Time-to-Digital Converters*, Analog Circuits and Signal Processing, DOI 10.1007/978-3-319-11842-0_4

resolution, the Vernier method [25, 95] was commonly used. However, the mismatch problem caused by process variation limits its effectiveness. Although calibration can be applied to compensate for the mismatch error [37], huge efforts are needed since each delay element in the TDC has to be tuned individually. Other methods to achieve sub-gate-delay resolution such as time amplification (TA) [49, 74], local passive interpolation (LPI) [38, 41], gated ring oscillator (GRO) [83], and successive approximation (SAR) [53] are vulnerable to transistor characteristic changing and temperature variation. Although calibration employing extensively extra feedback circuitries can be applied to maintain a nearly constant gate delay, its tuning ability is quite limited with regard to a wide temperature range of more than 200 °C and a multi-MGy total irradiation dose (TID). Moreover, the power consumption of delay-line-based TDCs increases linearly with the peak-to-peak amplitude of the input time signal, which is undesirable for a wide range measurement.

It is well known that, $\Delta\Sigma$ analog-to-digital converters (ADCs), which have been successfully implemented in analog-to-digital conversion for years, are highly immune to environmental noise and component mismatch. Some works brought the same principle into phase-domain data conversion, as in [91, 94], and achieved first-order and second-order noise shaping, respectively. However, these phase-domain $\Delta\Sigma$ ADCs are analog intensive approaches, and the phase information is converted back to the voltage domain. As the technology scales down, this becomes less attractive due to the difficulty of achieving high-gain and wide-bandwidth analog blocks under strongly reduced supply but relatively unchanged threshold voltage. Moreover, their performances are highly relying on the linearity of the front-end phase detector, which practically limits the dynamic range (DR) of the input time signal.

This work presents a radiation tolerant multistage noise shaping (MASH) delta-sigma ($\Delta\Sigma$) TDC [17]. It adopts the noise shaping concept, which can improve the effective resolution of a coarse quantizer, but requires no precisely matching analog components. The MASH TDC achieves better than 10 ps time resolution when a coarse quantization step of 16 ns is used, by shaping the quantization noise out from the interested baseband to high frequency. The on-chip-generated quantization reference clock has a frequency depending primarily on passive components, which shows intrinsic process, voltage, temperature (PVT) stability, and radiation tolerance. The data conversion is mainly being processed in the time domain, which benefits most from technology downscaling. Furthermore, the MASH TDC exhibits low-power nature owing to the employment of large quantization steps, which makes it suitable for applications pursuing a wide measurement range.

In this chapter, system level design and physical implementation of the MASH $\Delta\Sigma$ TDC will be discussed. First, the architecture of the third-order MASH $\Delta\Sigma$ TDC is investigated in Sect. 4.2. Detailed noise analysis of the TDC is given in Sect. 4.3, regarding mainly the jitter noise in the relaxation oscillator and the phase skew error introduced by the comparator delay. Section 4.4 shows the circuits implementation and measurement results of the first prototyping MASH $\Delta\Sigma$ TDC (chip I). In Sect. 4.5, the delay-line-assisted calibration technique is introduced to solve the phase skew problem. A demonstrated TDC (chip II) shows a similar time resolution

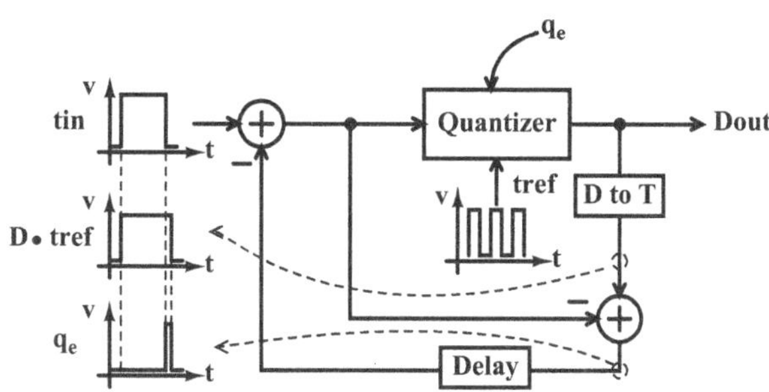

Fig. 4.1 Behavior model of the error-feedback-structure-based time-to-digital converter (*TDC*)

with chip I but the input DR has been increased five times, when the power consumption is reduced by more than half. Section 4.6 discusses the radiation assessment results at both low (1.2 kGy/h) and high (30 kGy/h) dose rate. A conclusion is drawn in Sect. 4.7.

4.2 Architecture of the 1-1-1 MASH $\Delta\Sigma$ TDC

4.2.1 The First-Order Error-Feedback TDC

The commonly used $\Delta\Sigma$ ADC structure which consists of integrators cannot be directly adopted by a TDC, due to the difficulty of realizing a time integrator. The error-feedback noise shaping structure introduced in the early 1960 is impractical for a $\Delta\Sigma$ ADC since its performance is limited by the inaccuracy of the analog subtractors. However, time subtraction can be easily realized by NAND/NOR operation, and moreover, the error-feedback structure does not require an explicit integrator in the loop. A first-order noise shaping TDC can be built in an error-feedback manner, as shown in Fig. 4.1. The input time signal t_{in} is first digitized by using a reference clock *tref*. The quantizer can simply be a counter, which is enabled/disabled by the input signal. A quantization error presents at the output when the input signal is not an integer times of the reference clock period. It can be reproduced by subtracting the input signal from the corresponding reference time of the digital output. A memory element is inserted in the feedback loop to preserve the quantization error before it is being subtracted from the next input signal.

However, directly preserving the quantization error in the time regime is still impossible with current technologies: the time information has to be converted into other intermediate physical quantities such as voltage or charge. A relaxation oscillator (Fig. 4.2a) can generate a clock by alternatively charging and discharging two capacitors. The phase of the clock corresponds to the voltage on each capacitor. The time can be measured by enabling the oscillator during the measurement interval and counting the number of periods of the generated clock. When the oscillation stops, the phase of the clock, which refers to the quantization error, can be stored on the capacitor as a residue voltage. First-order noise shaping can be achieved by

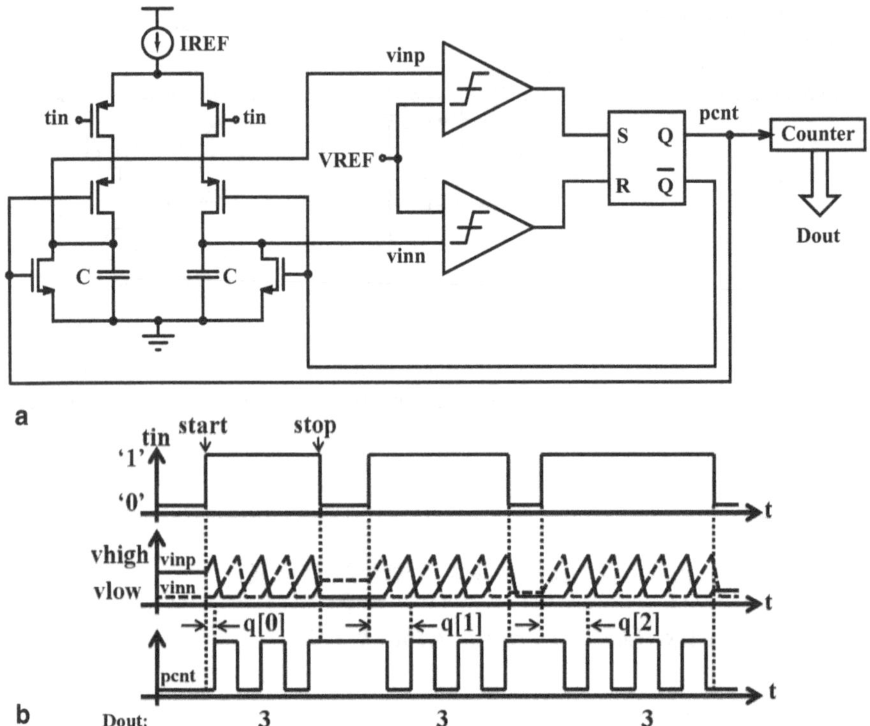

Fig. 4.2 **a** Schematic and **b** timing diagram of the relaxation oscillator-based first-order error-feedback time-to-digital converter (*TDC*)

forwarding this error information into the next measurement phase, hence canceling the low-frequency quantization noise. Principally, it is the same as the GRO [83], but the skew error caused by charge redistribution and path leakage during the start and stop of the oscillator is negligible here due to the large capacitance *C* and the controllable large charging slope (in the case of a not too slow counting clock, e.g., 62.5 MHz in this design).

The first-order error-feedback TDC works as follows: the time signal t_{in} controls a current to charge one of the two capacitors during its active phase. For instance, *vinp* starts rising when *vinn* stays at vlow, as illustrated in Fig. 4.2b. When *vinp* exceeds the threshold voltage *VREF*, the comparator output becomes "1." This reverses the state of the set–reset (SR) latch, and triggers the oscillation. The output of the oscillator is connected to a 4-bit counter. The final result in the counter is a digitized copy of the input signal with a large quantization error. After the stop signal arrives, the charging current is disconnected from the capacitors; the counter is first read out and then reset to 0. By preserving the residue voltage on the capacitor at the end of each measurement interval, the quantization error $q[k-1]$, which refers to the phase of the oscillator clock, is also preserved. The quantization error takes on a uniform distribution over the interval $[0, T_{\mathrm{d}})$, where T_{d} is the coarse quantization step size.

When the next measurement is initiated, the previous quantization error is subtracted from the next input, due to the fact that the counter is only driven by the rising edge of the clock. The overall quantization error introduced into this measurement can then be described as

$$q_{\text{err}}[k] = q[k] - q[k-1]. \tag{4.1}$$

If the quantization error from each measurement interval is adequately scrambled to random noise, this will results in first-order noise shaping of the quantization error.

4.2.2 High-Order Noise Shaping TDC

Similar to design of $\Delta\Sigma$ ADCs, the noise shaping concept can also be extended to higher orders. Recalling the structure of the classic single-loop second-order $\Delta\Sigma$ ADC, it replaces the quantizer in the first-order modulator with another $\Delta\Sigma$ ADC, which can further suppress the in-band quantization noise. However, in this configuration, integration of the input signal is unavoidable, which is difficult to perform and undesired in time domain operation. Instead, the $\Delta\Sigma$ ADC configured in a MASH architecture can obtain the same high-order noise shaping property but offer more freedom to choose a structure for each stage. For instance, a 1-1 MASH $\Delta\Sigma$ ADC, as shown in Fig. 4.3a, is built by cascading two identical single-loop first-order $\Delta\Sigma$ ADCs. A signal containing the quantization error from the previous stage is fed into the second stage. Two stages are combined together with the help of few additional digital processing blocks to achieve a second-order noise shaping.

In the case of building a 1-1 MASH $\Delta\Sigma$ TDC, two identical first-order error-feedback TDCs can be cascaded together, as shown in Fig. 4.3b. It is algebraically equivalent to the conventional 1-1 MASH $\Delta\Sigma$ ADC. The time signal which feeds into a following stage is generated by subtracting the quantization error from the input of the previous stage. This is done by taking the first rising edge of the counting clock as the new start signal, and keeping the same stop signal as the TDC's initial input. More details regarding this operation is described in the design of the time regenerator.

By cascading more error-feedback structures, a higher order noise shaping TDC can be formed. In this work, a third-order MASH $\Delta\Sigma$ TDC is demonstrated. The system architecture of a 1-1-1 MASH $\Delta\Sigma$ TDC is shown in Fig. 4.4. All three stages have the same structure and are followed by a digital processing block. Each stage works as a relaxation oscillator, controlled by the input time signal. The output of the 1-1-1 MASH TDC is given by

$$D_{out} = t_{in} + (1 - z^{-1})^3 \cdot q_{\text{err3}}, \tag{4.2}$$

where q_{err3} is the quantization error in the third stage. All digital blocks used for signal processing are synchronized by the falling edge of the input time signal. The

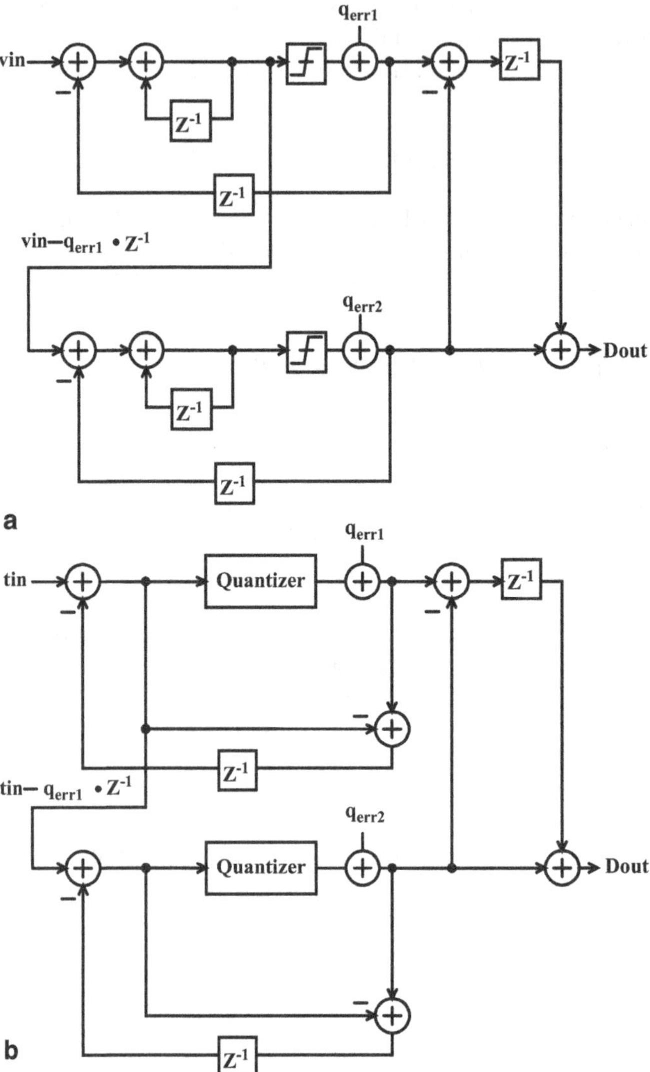

Fig. 4.3 Behavior models of **a** the classic 1-1 multistage noise shaping (*MASH*) $\Delta\Sigma$ analog-to-digital converter (*ADC*) and **b** the 1-1 MASH error-feedback time-to-digital converter (*TDC*)

theoretical *rms* value of the quantization noise power can be written as [59]:

$$q_{err_{rms}} = \frac{T_{\mathrm{d}}}{\sqrt{12}} \frac{\pi^L}{\sqrt{2L+1}} (OSR)^{-(L+1/2)}, \tag{4.3}$$

where *OSR* is the oversampling ratio, and *L* is the order of noise shaping function. An example TDC using a T_{d} as 16 ns, *OSR* as 25, and *L* as 3, will then ideally have

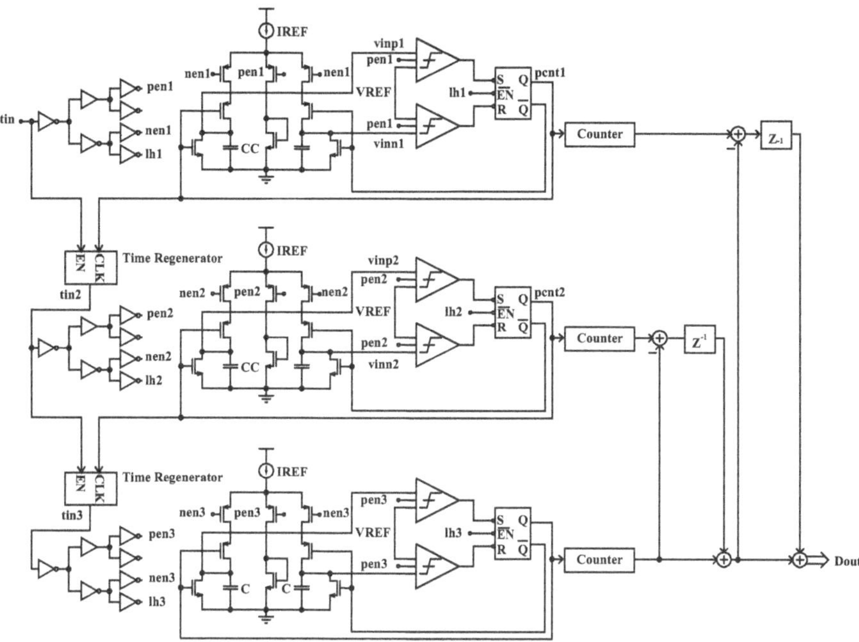

Fig. 4.4 System architecture of the 1-1-1 multistage noise shaping (*MASH*) $\Delta\Sigma$ time-to-digital converter (*TDC*)

rms quantization error of 0.7 ps. If a higher *OSR* of 250 is employed, and the other parameters remain the same, the theoretical *rms* quantization error can be reduced to only 0.2 fs, which is far below the physical noise floor of the TDC.

Although there is no physical limitation for all oscillator-based TDCs' input range, which can be easily extended by cycling the signal in the same TDC core, this is true only when the single-shot measurement is performed. If the input time signal has a nondirect current (DC) frequency, the measurement range of the TDC is in fact limited by the sampling rate. The reason is that, the stop signal of the current measurement has to come earlier than the next start signal, otherwise two input time signals will conflict. In the MASH $\Delta\Sigma$ TDC, there is also a special relation between the *OSR* and full-scale input range. The bandwidth of the input signal, BW, is set to 100 kHz in this design. The sampling clock of the TDC system is then equal to $2BW \cdot OSR$. Due to the input signal's timing nature, the peak-to-peak full-scale input signal amplitude T_{fspp} has to be smaller than one period of the sampling clock. For instance, when the *OSR* is 25, T_{fspp} cannot exceed 200 ns. This relation is shown in Fig. 4.5. When the ideal effective resolution of the MASH TDC improves with a higher *OSR*, the full-scale input range reduces. Therefore, when the TDC is being designed to achieve the targeted resolution, the signal-to-noise ratio (*SNR*) also needs to be optimized in order to achieve a wide input range. This design concept will be kept throughout the book.

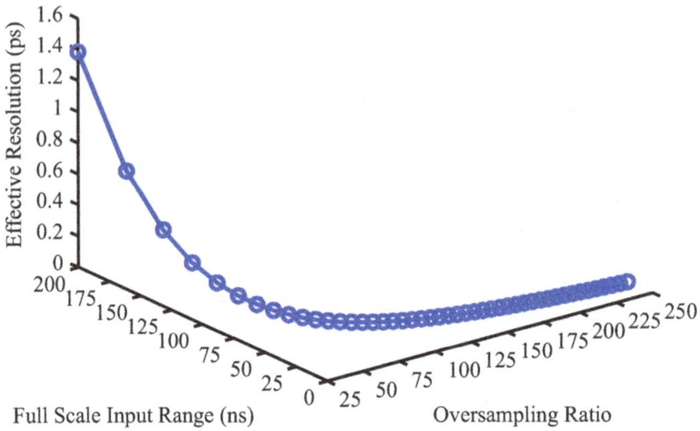

Fig. 4.5 Relation between oversampling ration (*OSR*), full-scale input range and the effective resolution.

4.3 Noise Analysis

When only the quantization noise is considered, the theoretical *SNR* of the 1-1-1 MASH $\Delta\Sigma$ TDC can be expressed as:

$$\begin{aligned} SNR &= 10\log\left(\frac{s_{rms}^2}{q_{\mathrm{err}_{rms}}^2}\right) \\ &= 20\log\left[\frac{\frac{(2^N-1)\cdot T_{\mathrm{d}}}{\sqrt{2}}}{\frac{T_{\mathrm{d}}}{\sqrt{12}}\frac{\pi^L}{\sqrt{2L+1}}(OSR)^{-(L+1/2)}}\right] \\ &= 10\log\left[6(2L+1)\right] - 10L + 20\left(L+\frac{1}{2}\right)\log(OSR) \\ &\quad + 20\log(2^N-1). \end{aligned} \tag{4.4}$$

In this design, *L* is set to 3, and *N*, the bits of the coarse quantizer, is set to 4. For an *OSR* of 25, the *SNR* is calculated as 107.7 dB, which turns out to an effective number of bit (*ENOB*) of 17.5 bits. However, the finest achievable resolution of the MASH TDC is practically limited by the jitter noise in the relaxation oscillator, phase skew caused by start/stop of the oscillator, and switching charge injection. Among them, the jitter noise and phase skew draw most attention, since they are at a much higher power level than the switching charge injection noise, which is actually negligible when the charging slope on the capacitors is sufficiently large. A detailed analysis of these two noise contributions is stated below. In this analysis, only the noise from the first stage is considered, since in the second and third stages, the circuit noise is reduced by first- and second-order noise shaping, respectively.

The noise analysis is based on the schematic shown in Fig. 4.2a. All the noise sources presented in the circuit can be modeled as shown in Fig. 4.6, where the fully

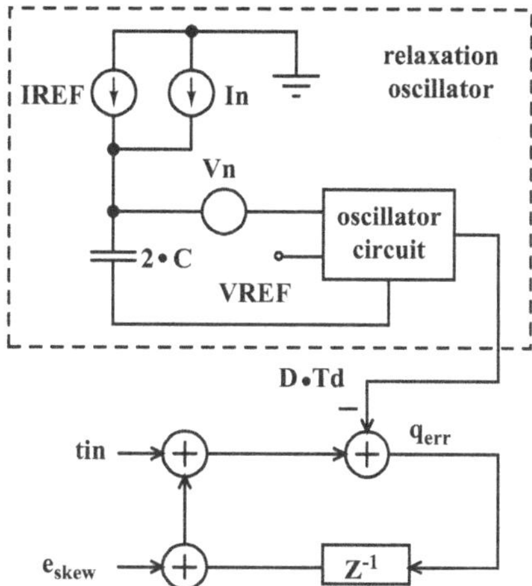

Fig. 4.6 Representation of noise in the relaxation oscillator based first-order $\Delta\Sigma$ (TDC)

symmetric relaxation oscillator is presented in the half-circuit mode. Then we have

$$q_{\text{err}} = tin + e_{\text{skew}} - D{\cdot}T_{\text{d}}, \tag{4.5}$$

$$T_{\text{d}} = \frac{(VREF + Vn){\cdot}2C}{IREF + In}, \tag{4.6}$$

where e_{skew} is the phase skew error caused by the comparator delay, D is the output code, T_{d} is the oscillation period, Vn is the equivalent input noise generator which represents the noise in the comparator, and In is the noise in the current source. The delay of the comparator is not included in the oscillation period, since it only adds a DC offset to the oscillation period. And its noise contribution to the system's performance is modeled as e_{skew}.

In and Vn are two primary contributors to the relaxation oscillator's jitter. The latter is generally the dominant cause of jitter, due to its much larger contributing bandwidth [3, 32]. Thus, the output code of the first-order TDC can be described as:

$$\begin{aligned} D &= \frac{tin + e_{\text{skew}} - q_{\text{err}}{\cdot}(1 - z^{-1})}{\frac{(VREF + Vn){\cdot}2C}{IREF}} \\ &= \frac{tin + e_{\text{skew}} - q_{\text{err}}{\cdot}(1 - z^{-1})}{\frac{2(VREF + Vn)}{S}}, \end{aligned} \tag{4.7}$$

where S is the charging slop of the capacitor. Therefore, apart from the noise-shaped quantization error, two additional noises also appear at the TDC's output. In order to derive the design specification of the TDC more accurately, it is important to know the impact of those two noise sources on the TDC's overall performance. Quantitative calculations of the TDC's *SNR* determined by the oscillator jitter and phase skew error, respectively, are stated below.

4.3.1 Timing Jitter

Assuming the timing jitter is the dominant noise in the TDC, the ideal *SNR* of the TDC can then be described as

$$\begin{aligned} SNR &= 10\log\left(\frac{T_{\mathrm{fs}}^2/2}{\sigma_{\Delta T_{\mathrm{OSC}}}^2/OSR}\right) \\ &= 6.02\cdot ENOB + 1.76, \end{aligned} \tag{4.8}$$

where T_{fs} is the full-scale input level, which is half of T_{fspp}, and $\sigma_{\Delta T_{\mathrm{OSC}}}$ is the *rms* jitter. Therefore, for an *OSR* of 25, in order to achieve an *ENOB* of 14 bits, the *rms* jitter needs to be smaller than 18 ps. Using a formula derived in [32]

$$PN(f_{\mathrm{m}}) = \frac{f_{\mathrm{OSC}}}{f_{\mathrm{m}}^2}\cdot\left(\frac{\sigma_{\Delta T_{\mathrm{OSC}}}}{T_{\mathrm{OSC}}}\right)^2, \tag{4.9}$$

where $PN(f_{\mathrm{m}})$ is the phase noise, f_{m} is the carrier offset frequency, and f_{OSC} is the oscillating frequency (62.5 MHz in this case). To fulfill the *rms* jitter requirement derived above, the oscillator needs to show a phase noise of better than −81.1 dBc/Hz at 100 kHz offset frequency. By substituting (4.9) into (4.8), we can get

$$SNR = 10\log\left(\frac{f_{\mathrm{OSC}}^3}{8\cdot PN(f_{\mathrm{m}})\cdot f_{\mathrm{m}}^4\cdot OSR}\right). \tag{4.10}$$

The relation between the *SNR* and jitter can also be predicted from a simulation of the MASH TDC behavioral model. The 1-1-1 MASH TDC was modeled in Simulink, which has the same operation behavior as in the real circuits. An additive white noise source is added on the reference voltage, which gives an estimation of the timing jitter in the oscillator. The result is shown in Fig. 4.7a. With a higher *OSR*, the *SNR* drops due to decrease of the full-scale input range as explained in Sect. 4.2. But it has better tolerance to the timing jitter, since the in-band jitter noise power can be reduced by oversampling. For an RC relaxation oscillator, which has a frequency of 62.5 MHz, the minimum achievable phase noise is approximately −100 dBc/Hz at 100 kHz offset frequency [57]. It means that the highest *SNR* can be achieved by the MASH TDC is 110 dB, when only the jitter noise in considered.

4.3.2 Phase Skew

The phase skew, e_{skew}, occurs only when the oscillator needs to be started and stopped. When the relaxation oscillator is turned off, the comparator state may not be perfectly preserved due to the hysteresis. This will introduce extra noise into the preserved quantization error, and it can only be suppressed by oversampling. The phase skew occurs when the stop signal arrives during the time the comparator enters

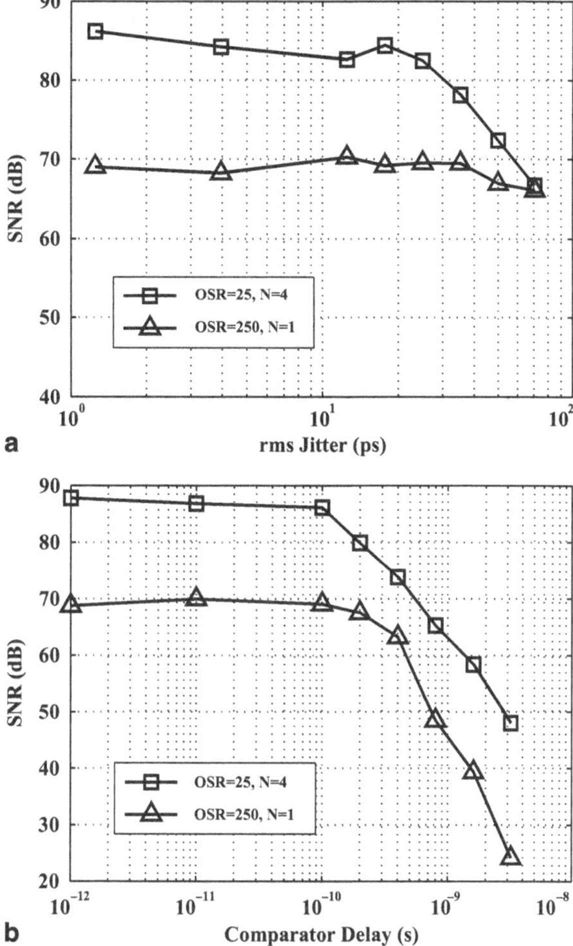

Fig. 4.7 Simulated signal-to-noise ratio (*SNR*) versus **a** timing jitter and **b** comparator delay

its state-reversing phase, when *vinn* or *vinp* just exceeds *VREF*. A delay always exists before the comparator can make a final decision according to its input change. It is impractical to save all the intermediate states of the comparator when the system enters idle. So the output of the comparator will continue rising till its final state "1," even if the oscillation has been stopped. Therefore, when the next start signal arrives, the SR latch will immediately reverse its state, and alternates the capacitor being charged. This will result in a change in the counting clock period and introduce extra phase skew error to the preserved quantization residue time. This error is shown by red lines in Fig. 4.8.

The relation between the *SNR* and the comparator delay, as shown in Fig. 4.7b, is predicted by simulating the same Simulink model used in timing jitter estimation. Mathematically, when the stop signal arrives, the time domain representative of the voltage on the capacitor follows a uniform distribution over the interval $[0, T_{OSC}/2)$.

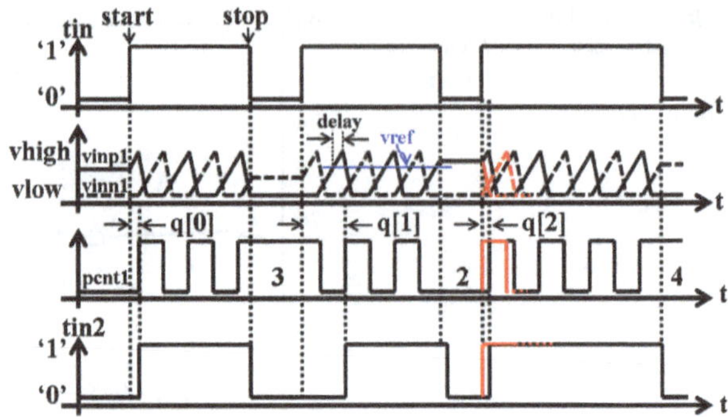

Fig. 4.8 Illustration of the phase skew introduced by the comparator delay

When this voltage is aligning between $T_{OSC}/2 - t_{cmp}$ and $T_{OSC}/2$, where t_{cmp} is the comparator delay, the random skew error will occur. The mean value of this error equals $t_{cmp}/2$, and the probability density function of the error is

$$f(e_{skew}) = \frac{1}{T_{OSC}/2}. \tag{4.11}$$

Therefore, the variance of the phase skew error can be described as

$$\begin{aligned} \sigma^2_{e_{skew}} &= \int_0^{t_{cmp}} \left(e_{skew} - \frac{t_{cmp}}{2}\right)^2 \cdot \frac{2}{T_{OSC}} de_{skew} \\ &= \frac{t^3_{cmp}}{6 \cdot T_{OSC}}. \end{aligned} \tag{4.12}$$

Similar to (4.8), when only the phase skew error is taken into account, the theoretical *SNR* of the MASH TDC can be calculated as:

$$SNR = 10 \log \left(\frac{T^2_{fs}/2}{\sigma^2_{e_{skew}}/OSR} \right). \tag{4.13}$$

Thus, for an *OSR* of 25, if the comparator has a delay of 1 ns, the *SNR* will be only 70.8 dB. When increasing the *OSR*, the skew error can be reduced by the same order. However, the full-scale input is decreasing at the same time. Eventually, the *SNR* of the TDC remains almost unchanged. But the time resolution ($T_{fs}/2^{ENOB} - 1$) can be improved at the cost of a smaller input time range. Comparred to the quantization noise and timing jitter, the phase skew error becomes the dominant noise in the MASH TDC. In order to achieve a higher *SNR*, either the comparator delay has to be limited to a small amount or calibration needs to be applied.

Other path delays, such as the one in the SR latch and the gate delay of the oscillator switches, have less impact on the performance of the TDC. This is due to the fact that the input signals to the SR latch and oscillator switches are already rail-to-rail signals with sharp rising/falling edges, and thus those delays can be made sufficiently small, e.g., < 100 ps.

4.4 Chip I: First Prototyping of the MASH $\Delta\Sigma$ TDC

4.4.1 Circuit Description

The main circuit blocks in the 1-1-1 MASH $\Delta\Sigma$ TDC are the relaxation oscillator, the counter, the time regenerator, and digital processing units. Among them, the performance of the relaxation oscillator determines the highest achievable *SNR* of the TDC. The conventional two-capacitor relaxation oscillator structure [28] is used in this design, due to its simplicity of controlling. The specification of the relaxation oscillator is derived first according to the noise analysis presented in Sect. 4.3, and then a trade-off has been made between the time resolution and power consumption.

4.4.1.1 Relaxation Oscillator

The on-chip relaxation oscillator provides the reference clock for the $\Delta\Sigma$ TDC, whose frequency therefore needs to be stable over process and temperature. The period of the relaxation oscillator can be expressed as $(VREF{\cdot}2C)/IREF{+}2t_{\mathrm{cmp}}$. If the comparator delay t_{cmp} is small enough comparing to the whole clock period, the oscillator frequency becomes $IREF/(VREF{\cdot}2C)$. By correlating *VREF* and *IREF* as $VREF = IREF{\cdot}R$, its frequency becomes only depending on passive components, which is $1/(2{\cdot}RC)$. Thus, it exhibits inherent PVT variation tolerance and the matching between stages is better than for its MASH ADC counterparts. Furthermore, the matching between Q and $\bar{\mathrm{Q}}$ path is not important, since only the whole period of the oscillator clock is used as one quantization step, regardless of the duty cycle. In addition to the jitter noise in the relaxation oscillator, charge injection when turning on/off the input control switches and their associated leakage current also contributes to the overall noise level. In the conversion to time noise, the local noise voltage is divided by the charging slope of the capacitor. Thus, a larger slope is mostly desirable. In this design, $IREF = 50\,\mu$A, $VREF = 650$ mV, and $C = 0.64$ pF, which gives a charging slope of 80 μV/ps. According to simulation, the relaxation oscillator has a phase noise of -87 dBc/Hz at 100 kHz offset frequency, which is adequate to achieve better than 14 bits *ENOB*.

In order to limit the impact of the phase skew error introduced by the large comparator delay on the overall TDC's performance, fast comparators must be used. Figure 4.9 shows the schematic of the threshold-detection comparator used in the relaxation oscillator. It is built in a multistage structure for a high-speed consideration. Each of the first three stages has a gain of 10 dB and consumes 40 μA current. The last stage provides a higher gain of 20 dB with a power consumption of 80 μA. Input differential pairs of all stages formed by transistors M1a, M1b, M2a, M2b, M3a, M3b, M4a, and M4b are optimized for both matching and speed. Two comparators in each stage, controlled by the enabling signal pwd, are turned off alternately to save power, when its connected capacitor is not being charged. The switch M5 in the last stage, controlled by *EN*, is added to further reduce the effective phase skew caused by the comparator delay. Overall the comparator consumes 200 μA and has an effective

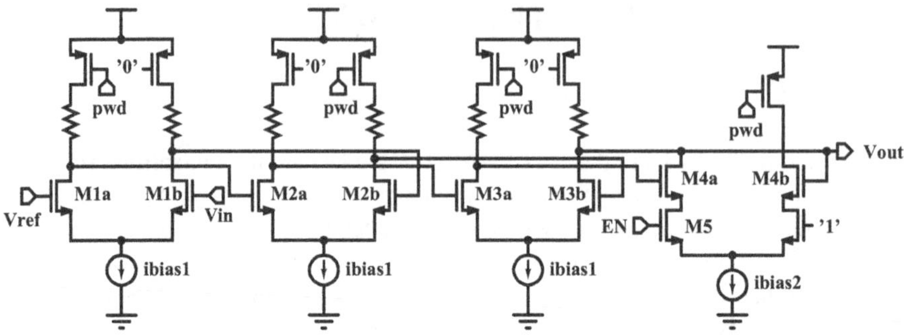

Fig. 4.9 Schematic of the four-stage threshold detection comparator

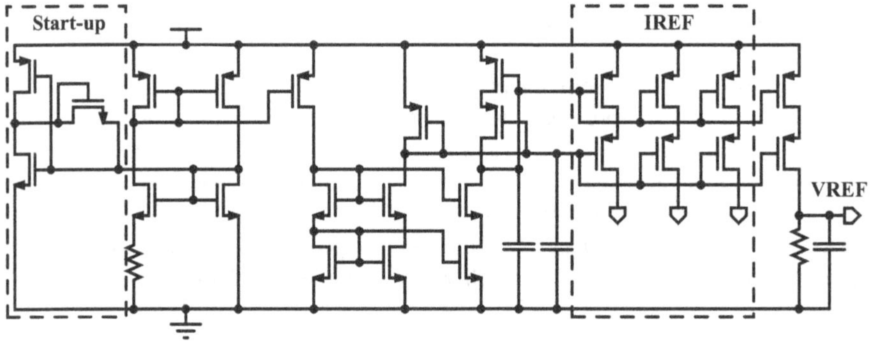

Fig. 4.10 Schematic of the current reference

delay of 800 ps, which is capable to achieve an signal-to-noise plus distortion ratio (SNDR) of around 65 dB when the *OSR* is 25, according to simulations. A higher SNDR is achievable by employing a faster comparator. However, it consumes too much power to implement a high-speed continuous-time comparator in this technology. A constant- g_m biasing circuit (Fig. 4.10) is used to provide biasing current for the comparator. It adjusts the biasing current adaptively according to the threshold voltage variation of the MOS transistor, and keeps the transconductance constant. It also generates the charging current *IREF* for each stage and the reference voltage *VREF* for the comparator.

4.4.1.2 Time Regenerator

The time regenerator produces the input time signal for a following stage. As explained earlier, it generates a timing pulse whose rising edge is aligned with the first rising edge of the counting clock, and the falling edge comes when the stop signal arrives. As illustrated in Fig. 4.11a, $t_{in}[0]$ is first digitized, resulting a quantization error q_{err}. This error is subtracted from the second input signal $t_{in}[1]$ to generate the

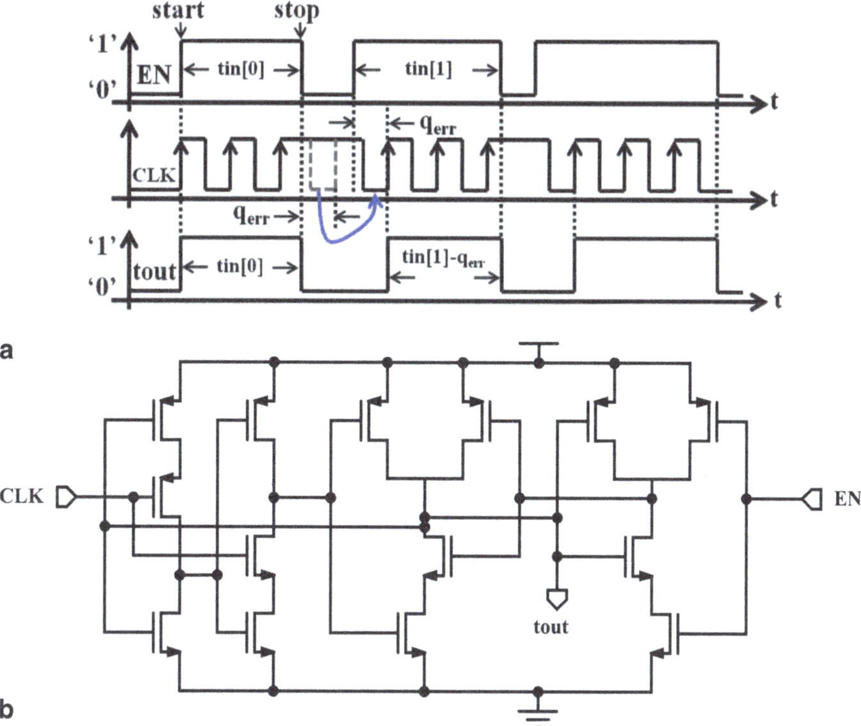

Fig. 4.11 **a** Timing diagram and **b** schematic of the time regenerator

input signal for the second stage. This can be achieved by using an RS latch. In order to inactivate the second, the third,... rising edges of the counting clock, a D flip flop is placed at the clk path before the RS latch, as shown in Fig. 4.11b.

4.4.1.3 Counter

A 4-bit counter is used as the quantizer in the MASH TDC, as shown in Fig. 4.12. Since there is only low-speed operation involved, the counter is built in an asynchronous style. The timing between enable signal *EN*, reset signal rst, clock signal clk, and latch signal *SP* is very crucial for appropriate operation of the counter. The last transition of the counting clock clk may come later than the stop signal, which locates at the falling edge of *EN*, due to the comparator delay. Therefore, the sampling clock of the output register should have enough delay to *EN* to wait until the counter is stable. And naturally, the reset signal of the counter has to be later than *SP*. This time relation is guaranteed by using asymmetrical inverters to buffer the enable signal *EN*, as illustrated in Fig. 4.12.

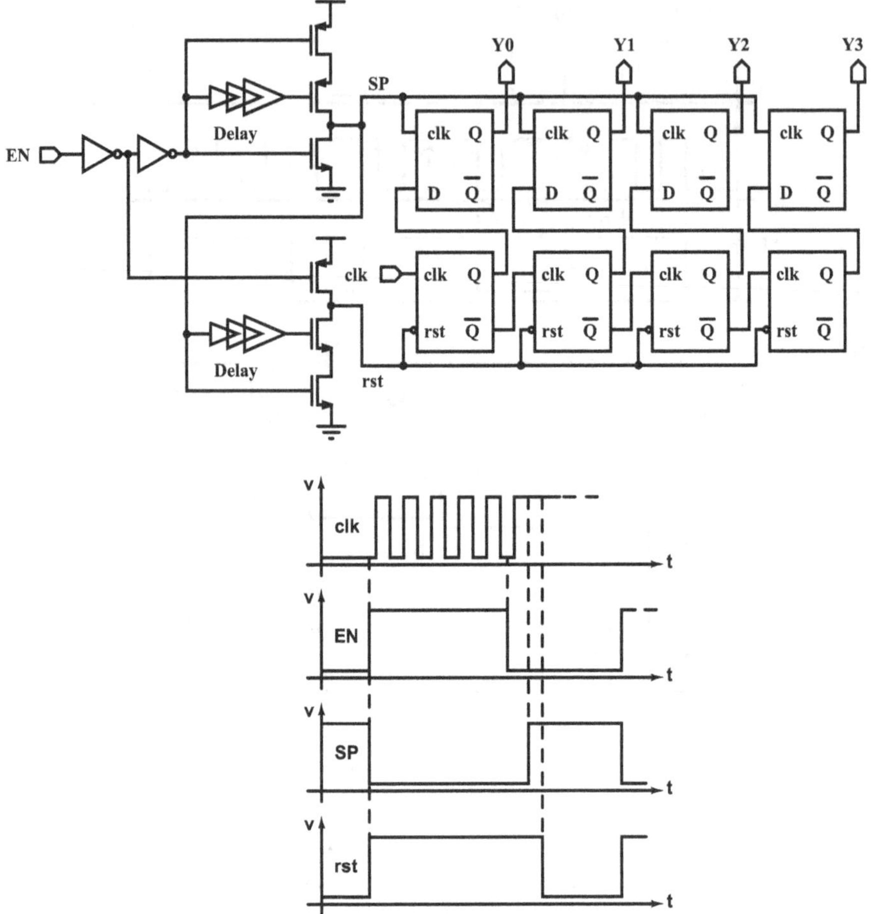

Fig. 4.12 Structure of the four-bit counter

4.4.2 Experimental Results

The 1-1-1 MASH $\Delta\Sigma$ TDC is implemented in 0.13 μm CMOS. It consumes 1.7 mW from a 1.2 V supply. The die photo of the TDC is shown in Fig. 4.13. A time domain sine wave signal is employed to evaluate the dynamic performance of the TDC. The timing pulse containing the start and stop signal at the rising and falling edge, respectively, is directly taken from the signal generator. The signal source generates a pulse width modulated (PWM) square wave, and the width of each pulse (corresponding to the time period) is modulated by a sine signal whose frequency is between DC to 100 kHz. In order to measure a time system with better than 10 ps resolution, it is not possible to use any digital-fashion arbitrary waveform generators, which usually have a signal rate of less than 50 Gsamp/s. Instead, a continuous-time

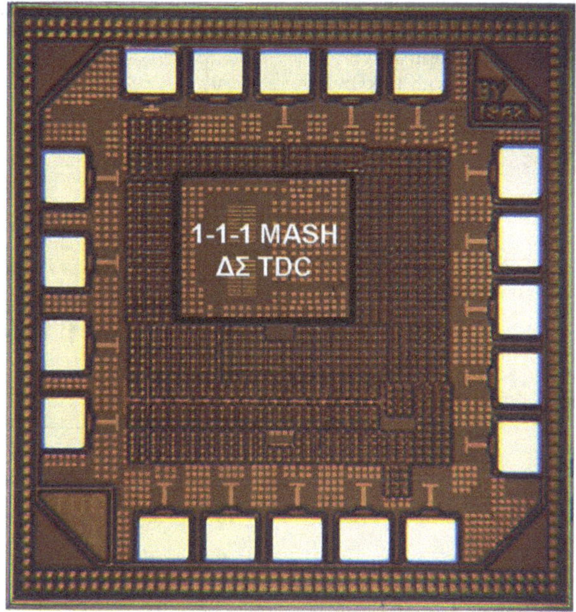

Fig. 4.13 Die photo of the multistage noise shaping (MASH) $\Delta\Sigma$

sine wave signal generator with infinite time resolution, the Agilent E8257D, is adopted in this measurement. The frequency-modulated sine wave is shaped into a PWM square wave by feeding the signal into a second external-triggered pulse generator (Agilent 8131A). In a real application, an edge combiner can be added on-chip to generate the timing pulse for the TDC core. Although the edge combiner might fail when the signal width becomes considerably small (e.g., <10 ps), it's not the case for most applications including the TOF rangefinder (with 1–30 m detection range). The dynamic measurement was performed with this PWM signal which has a DC bias of 100 ns down to 10 ns.

The conversion rate of the TDC can be varied from 5 to 50 MHz. For a signal bandwidth of 100 kHz, it turns to an *OSR* of 25–250. The TDC is configured first in a low conversion rate (5 MHz) mode. The full-scale input range determined by the conversion rate is then 200 ns. An 18 kHz −3 dBFS PWM signal is applied to the input of the TDC. The output spectrum and waveform are shown in Fig. 4.14. It shows an SNDR of 60.3 dB. The resolution of the TDC is mainly limited by the phase skew error introduced by the comparator delay. It can be reduced by increasing the *OSR*. In the second measurement, the *OSR* is increased to 250. This requires a higher conversion rate, which is now 50 MHz. The input full-scale range is hence reduced to 20 ns. The power spectrum and waveform of the TDC output, as shown in Fig. 4.15, are carried out with a 22 kHz −40 dBFS input, which has a peak-to-peak amplitude of 200 ps. An *ENOB* of 11 bits and an effective resolution of 5.6 ps are achieved, respectively.

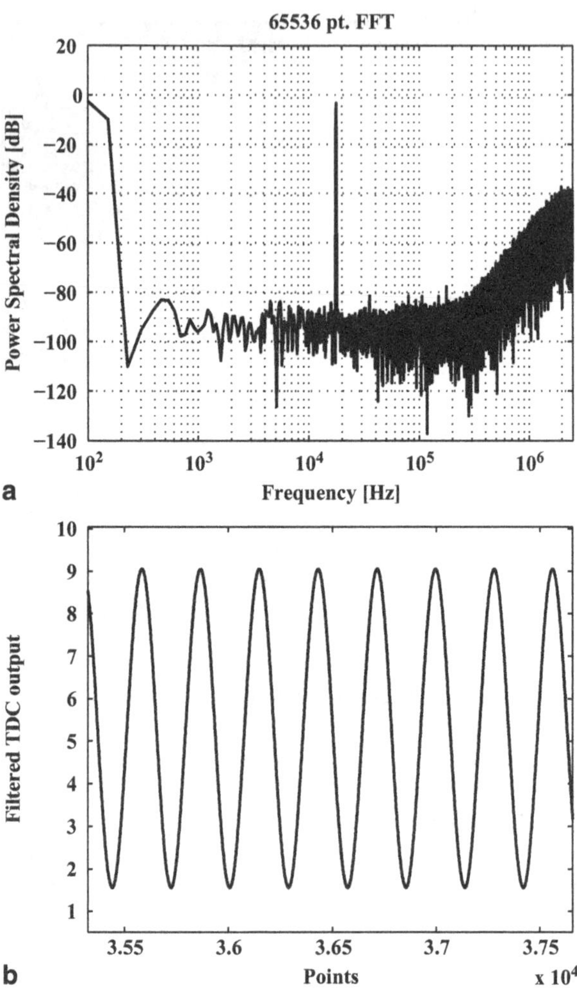

Fig. 4.14 **a** Measured power spectral density (PSD) and **b** output waveform after 100 kHz LPF, with 18 kHz −3 dBFS input ($OSR = 25$)

One should notice that, although increasing the *OSR* can reduce the phase skew error in the comparator, and further improve the TDC's SNDR, the effective quantization bits of the counter are also reduced at the same time, since the quantization step size T_{OSC} is remaining constant. Consequently, the full-scale SNDR and *ENOB* of the TDC will stay nearly unchanged regardless of the value of *OSR*. Therefore, a trade-off between time resolution and input range exists in this TDC. For instance, when a millimeter accuracy is targeted by a TOF rangefinder, the TDC needs to have 6.7 ps resolution. When the signal band of interests is fixed to 100 kHz, this MASH TDC can only offer a measurement range of 20 ns, which corresponds to a detection range up to 3 m in distance. In order to achieve the same time resolution but a much wider input range, the *SNR* of the TDC has to be increased when a smaller *OSR* is

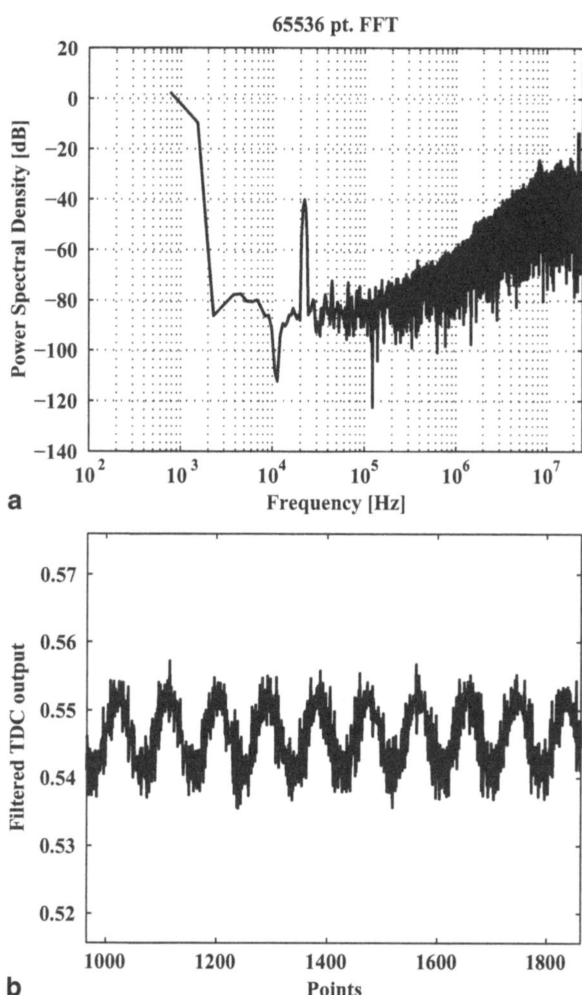

Fig. 4.15 **a** Measured PSD and **b** output waveform after 100 kHz LPF, with 22 kHz −40 dBFS input ($OSR = 250$)

adopted. This can be achieved by reducing the comparator delay with more power consumption, or alternatively, by applying calibration to the phase skew error, as will be discussed in Sect. 4.5.

The system is also compatible with other *OSR* values, such as 50 and 100. Figure 4.16 shows the DR of the TDC, which is 68 dB. Note that, there is no apparent drop in SNDR when the input level is close to full scale, since in a TDC system, the maximum input amplitude is physically limited only by the depth of the counter, which can be easily extended to avoid any overloading of the system.

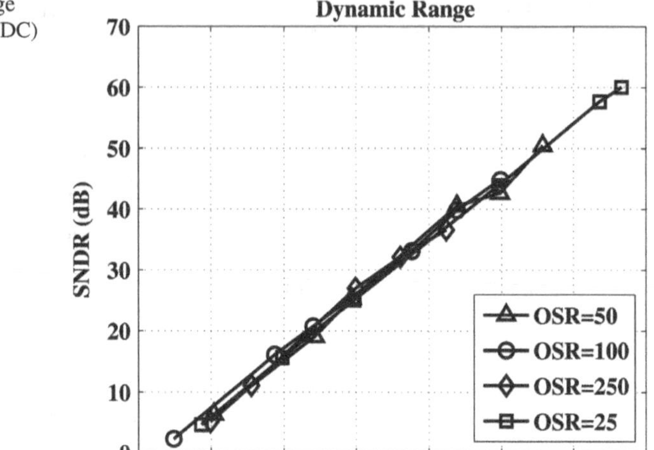

Fig. 4.16 Dynamic range (DR) of the (MASH) (TDC)

4.5 Chip II: The MASH $\Delta\Sigma$ TDC with Delay-Line-Assisted Calibration

4.5.1 Delay-Line-Assisted Calibration

As explained in Sect. 4.3, the main origin of noise in the MASH TDC is the phase skew introduced by the comparator delay. According to simulation results, the delay of the comparator has to be limited to 200 ps in order not to degrade the SNDR of the 1-1-1 MASH TDC, when a moderate *OSR* of 50 is adopted. A threshold-detection comparator at that high speed consumes huge power, and for some old technologies, it is even unrealistic to achieve that delay. Fortunately, this skew error can be calibrated by using a coarse delay line [16].

One stage of the 1-1-1 MASH $\Delta\Sigma$ TDC with delay-line-assisted calibration (TDC-CAL) is shown in Fig. 4.17. As one can see from it, the RS latch is controlled by the output of the calibration unit *EN* rather than t_{in}, but both have the same "stop" edge. *lh*, the complementary signal of t_{in}, is sent to the delay line, whose detailed structure is also shown in Fig. 4.17. Each delay cell has a delay of 150 ps. The delay line calibration unit becomes active only after the stop signal arrives. When the comparator's output becomes "1" during the inactive phase of the TDC, the state of each delay cell will be sampled by its connected arbiter.

For example, the stop signal arrives when *vinp1* rises above *vref*, as illustrated in Fig. 4.18. Therefore, the left comparator has entered its state-reversing phase. After tskew time, which is the resulting phase skew error, *cmp1* becomes "1." When *lh* has been sent to the delay line as the start signal, *cmp1* serves as the stop signal to sample the state of the delay line. If *lh* has passed through three delay cells, the states of all arbiters will be "1110...0" The switch connected to the third delay cell

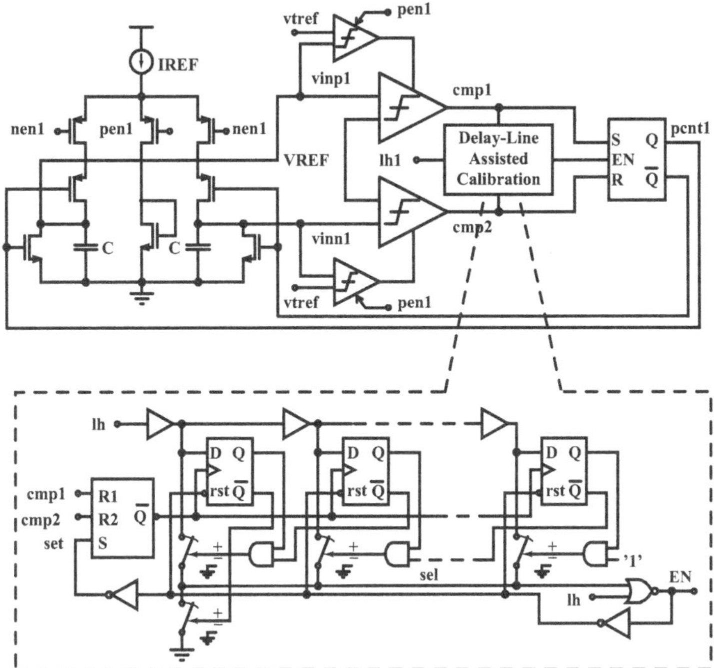

Fig. 4.17 One stage of the (MASH) (TDC) with delay-line (ten delay cells) assisted calibration

will also be closed, which charges *sel* to "1." *sel* will be discharged only after the next start signal has also passed through the same delay cells as the stop signal, then activate *EN*. The resulting waveform on the capacitor is shown in Fig. 4.18 by blue lines. In this way, the phase skew caused by the comparator delay is compensated at an accuracy of 150 ps. Since the rising edges of *lh* and *sel* have always passed through the same delay cells, the matching between those cells in the calibration unit becomes less important.

However, one problem might limit the effectiveness of the delay-line calibration method, which is the input-dependent delay of the comparator. When the input differential voltage to the comparator is very small, the comparator exhibits a much larger delay. And the relationship between the input voltage and comparator delay is nonlinear. As soon as the input differential voltage exceeds a certain level, which is 10 mV in this case, the difference in delay becomes negligible compared to the minimum detectable delay time by the calibration unit.

In order to avoid this nonlinear behavior of the comparator, an arming comparator has also been added to control each main comparator. It has a different reference voltage *vtref* which is slightly higher than that for the main comparator (10 mV higher in this case). When the stop signal arrives, it compares the residue voltage with *vtref*. There are three different cases: (1) the residue voltage is smaller than *VREF*, (2) the residue voltage is larger than *vtref*, and (3) the residue voltage is larger

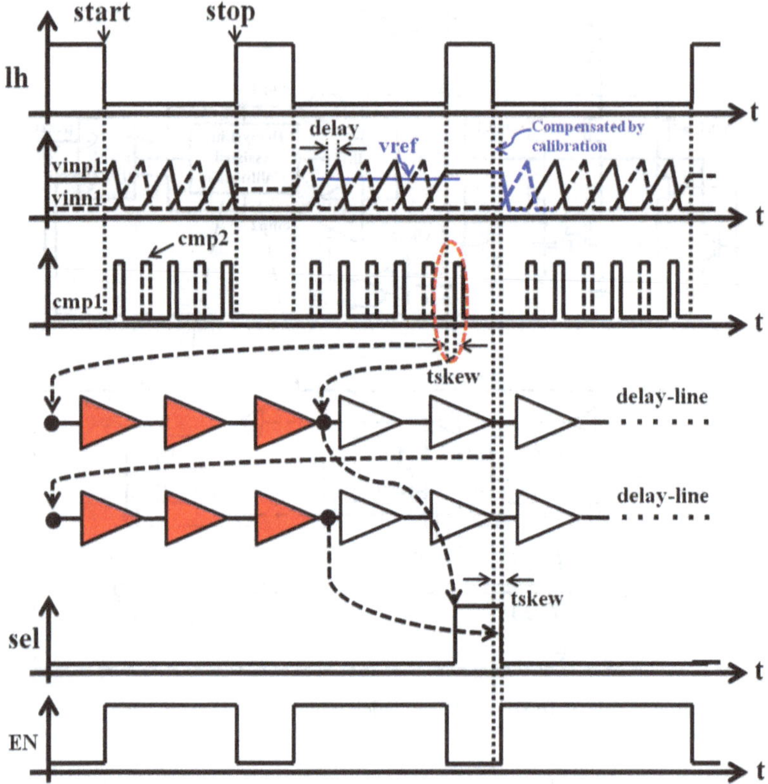

Fig. 4.18 Illustration of the delay-line-assisted calibration principle

than *VREF*, but smaller than *vtref*. In case 1, the main comparator is not its state-reversing phase, therefore no skew error will occur. In case 2, the main comparator has an input-independent delay, and that can be calibrated by the calibration unit. In case 3, the output signal of the arming comparator will disable the input stage of the main comparator. The state of the main comparator will be held as "0" even when the actual input *vinn* or *vinp* exceeds *VREF*. This avoids the comparator's nonlinear-large-delay state caused by very small input. Meanwhile, the introduced phase skew due to this operation is tolerable by the TDC.

4.5.2 Physical Implementation

The overall architecture of the 1-1-1 MASH TDC with delay-line-assisted calibration is shown in Fig. 4.19. The TDC-CAL uses the same digital processing block as in chip-I, as well as the counter and time regenerator circuits. In order to improve

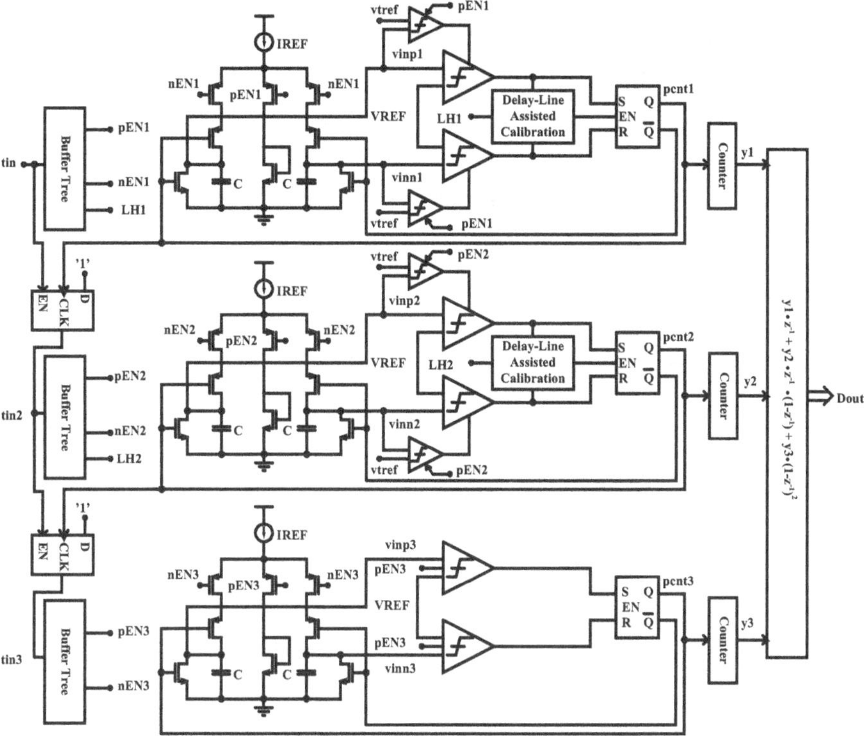

Fig. 4.19 Architecture of a 1-1-1 (MASH) (TDC) with delay-line assisted calibration

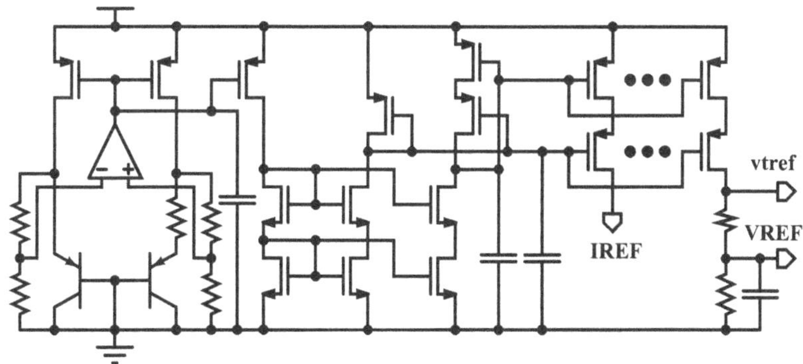

Fig. 4.20 Schematic of the bandgap reference

the TDC's temperature stability, an on-chip bandgap current reference [51] is implemented to provide 50 μA charge current for each stage, as shown in Fig. 4.20. Figure 4.21 shows the schematic of the main comparator in the TDC. Unlike the one adopted in chip-I, it is not necessary for this comparator to be high speed. Therefore,

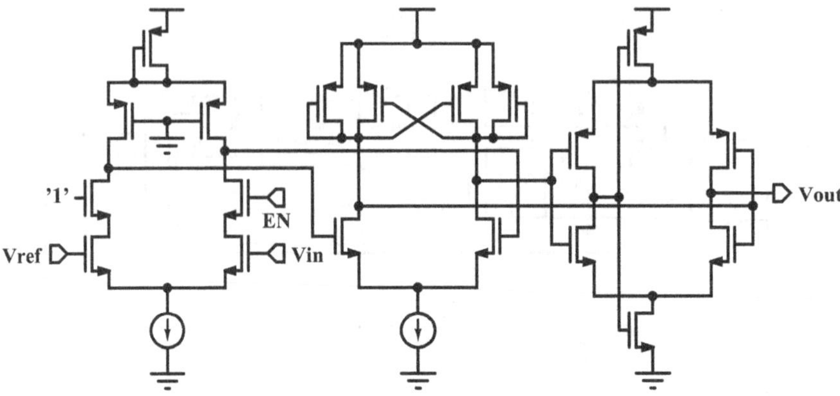

Fig. 4.21 Schematic of the comparator

a low-power threshold-detection comparator with moderate speed is implemented in the TDC-CAL. The comparator has three stages. The input stage is a low-gain high-bandwidth preamplifier. It tracks the input and makes fast decision as soon as the threshold condition is reached. The input differential voltage is then amplified and fed into the latch stage. A self-biased differential amplifier [12] is implemented as the output buffer stage. It consumes nearly zero static current, but has the ability to source and sink large currents. One comparator draws 30 μA quiescent current from the supply, and has a total delay of around 1.5 ns. A larger comparator delay is also tolerable by the design; however, this requires a larger detection range of the delay-line calibration unit, which means more circuits complexity and larger area. The arming comparator consists of a preamplifier and a clocked dynamic latch. It has the same preamplifier stage as the main comparator in order to minimize the mismatch between the two comparators.

4.5.3 Measurement Results

The TDC-CAL is implemented in 0.13 μm CMOS. The chip consumes 0.7 mW from a 1.2 V supply, and occupies 0.08 mm^2 area (core). The photo of the die is shown in Fig. 4.22. The same PWM signal described in the previous measurement setup is again used here to evaluate the performance of the TDC-CAL. The rising edge of the PWM pulse represents the start signal, where the stop signal is located at the falling edge. The carrier frequency is set to 10 MHz, which turns to a full-scale peak-to-peak input range of 100 ns. Then for a bandwidth of 100 kHz, the *OSR* is 50. Figure 4.23 shows the output spectrum of the MASH TDC with an 18 kHz 10 ns (−20 dBFS) peak-to-peak input. It shows an SNDR of 55.2 dB and 6 ps effective resolution with 100 kHz bandwidth. It can be clearly seen that, without calibration, the large phase skew caused by the comparator delay will introduce distortion and increase the baseband noise.

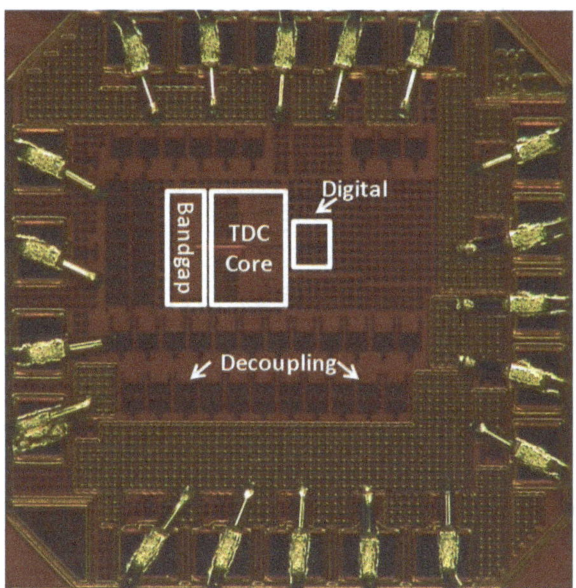

Fig. 4.22 Die photo of the (MASH) (TDC) with delay-line-assisted calibration

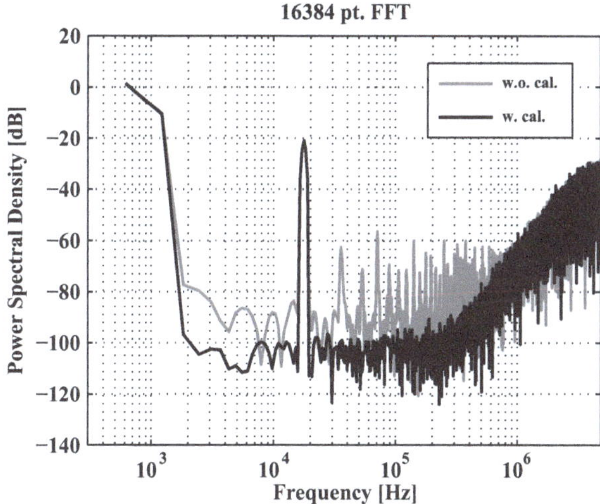

Fig. 4.23 Measured PSD with 18 kHz −20 dBFS input

The TDC-CAL has also been examined under different temperatures. As illustrated in Fig. 4.24, with a 41-ns DC input, the total shift of the TDC's output is less than ± 1.25 % over a wide temperature range of – 20 to 120 °C, which shows a temperature coefficient of 176 ppm/°C (7.2 ps/°C) without any calibration. From 25 to 120 °C, this value is only 76 ppm/°C. During the dynamic measurement, the SNDR of the TDC does not drop even when the temperature rises up to 100 °C.

The performance of two MASH TDCs is summarized in Table 4.1, and a comparison with state-of-the-art TDCs is also shown. Some report higher performances

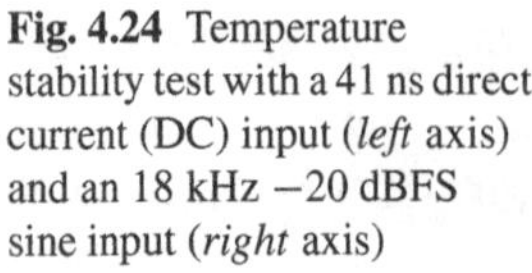

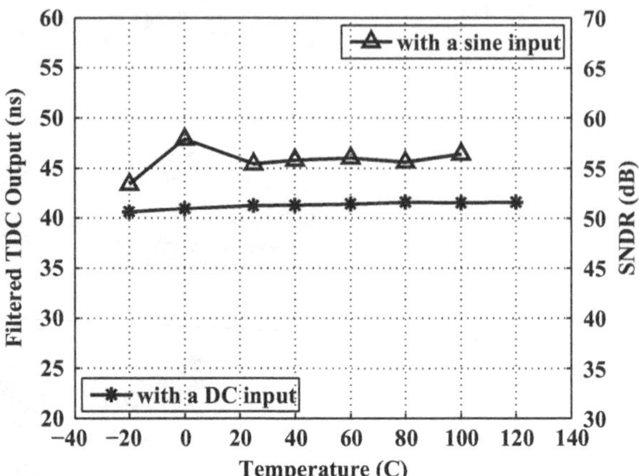

Fig. 4.24 Temperature stability test with a 41 ns direct current (DC) input (*left* axis) and an 18 kHz −20 dBFS sine input (*right* axis)

if only small (± 1 lower sideband, LSB) signals are applied, but in order to obtain a correct comparison only the full-scale data is used. Thanks to the MASH $\Delta\Sigma$ technique, the TDCs presented in this work achieve 5.6/6 ps effective resolution when only consuming 1.7/0.7 mW for a measurement range of 20/100 ns, at the cost of a smaller bandwidth (100 kHz in this work). Similar performances are achieved for different signal levels varying from 1 LSB to near full scale. In the MASH TDC with delay-line-assisted calibration, the resolution of the TDC can be further improved to sub-ps range by increasing the *OSR*, at compromise of the input range. Comparing to the GRO TDC [83], which demonstrates an effective resolution of 1 ps in 1 MHz bandwidth, but consumes 21 mW for an input of 12 ns, the MASH $\Delta\Sigma$ TDC with calibration consumes only 0.7 mW for an input of 100 ns. Although lower resolution and smaller bandwidth are obtained by the MASH TDC, the low-power feature and better tolerance to PVT variation made it a favorable choice for TOF measurement application.

4.6 Radiation Assessment of the MASH $\Delta\Sigma$ TDC

Radiation assessments have been performed at the Belgian Nuclear Research Centre, SCK·CEN. The low-dose-rate irradiation experiment was carried out with the "RITA" facility, where five TDC samples were irradiated with ^{60}Co gamma source. The measured dose rate was 1.2 kGy/h. The frequencies of the relaxation oscillators in five TDC samples are monitored continuously during the whole 130 h irradiation period. Figure 4.25 shows the real-time results obtained from a digital oscilloscope. The five frequencies differ from each other due to the die-to-die variation. From 0 to 160 kGy total ionizing dose (TID), no degradation on the frequencies has been found. Only small ripples exhibit, which are mainly caused by the jitter noise coupling into the cable (10 m long). This proves the radiation hardness of the RC relaxation oscillator.

Table 4.1 Comparison of state-of-the-art time-to-digital converters (TDCs) with similar specifications

	[49]	[38]	[83]	[53]	[95]	[94]	Chip-I	ChipII
Technique	TA	LPI	GRO	SAR	Vernier	Phase domain $\Delta\Sigma$	Time domain $\Delta\Sigma$	Time domain $\Delta\Sigma$
Sample rate (MS/s)	10	180	50	100	15	156	50	10
Resolution (ps)	1.25	4.7	6/1[a]	1.22	8	2.4	5.6	6
Bits	9	7	11	15	12	10	11	13
Measurement range (ns)	0.64	0.6	12	40	32	3.2	20	100
Power (mW)	3	3.6	21	33	7.5	2.1	1.7	0.7
Core area (mm^2)	0.6	0.02	0.04	1.2	0.26	0.12	0.11	0.08
CMOS	90 nm	90 nm	0.13 μm	0.35 μm	0.13 μm	90 nm	0.13 μm	0.13 μm

[a] 6 ps resolution is achieved with a full-scale input; 1 ps resolution is achieved when a small signal is applied

CMOS complementary metal–oxide–semiconductor, *TA* time amplification, *LPI* local passive interpolation, *GRO* gated-ring-oscillator, *SAR* successive approximation, *TDCs* time-to-digital converters

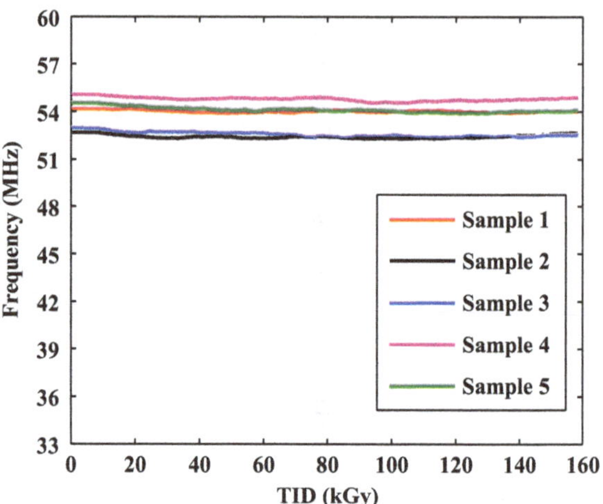

Fig. 4.25 Measured counting clock frequencies of five different samples. These measurements were obtained from the low-dose-rate experiment

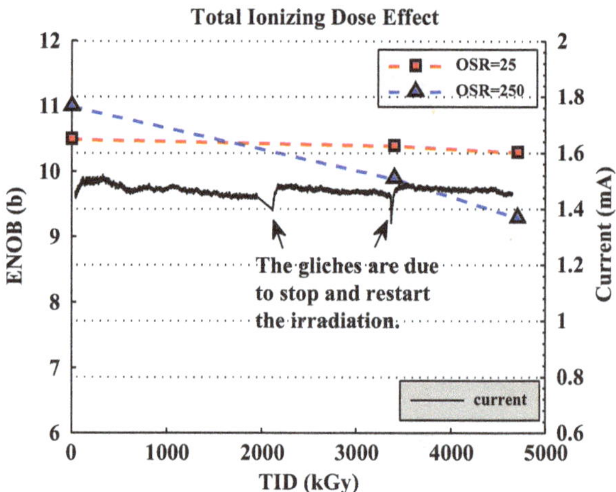

Fig. 4.26 Measured effective number of bits (*ENOB*) and total current consumption of the $\Delta\Sigma$ (TDC), obtained from the high-dose-rate experiment

An online dynamic measurement under a high-dose-rate radiation has also been done, proving the TDC's robustness. The experiment was carried out with the "Brigitte" facility, at a dose rate of 30 kGy/h. Figure 4.27 shows the setup of the experiment. The ceramic substrate with the bonded TDC (unpackaged) is mounted in a container, which is placed in the under water gamma irradiation facility. The substrate is connected to all measurement equipment by a cable with 10 m length. The online measurement results are shown in Fig. 4.26. When the system is working in the high conversion rate (50 MHz) mode, the *ENOB* of the TDC drops only 1 bit after a total dose of 3.4 MGy. This means that a resolution of 10.5 ps can still be achieved. The TDC works functionally till at least 5 MGy. For the TDC working in the low conversion rate (5 MHz) mode, it shows less drop in performance, which can be clearly seen from Fig. 4.26. This is because the timing noise converted from the

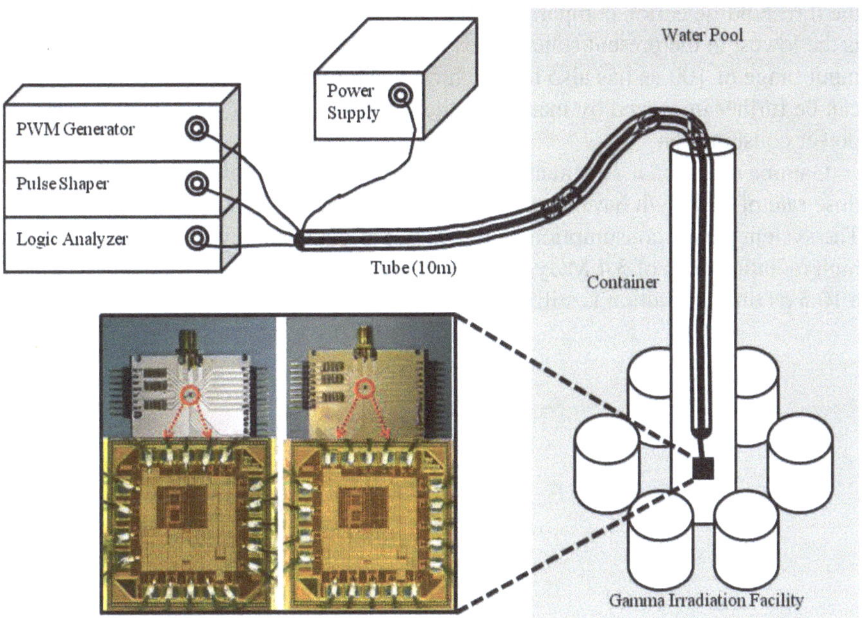

Fig. 4.27 Setup of the high-dose-rate gamma irradiation experiment. Photos of the substrate and the die before (*left*) and after (*right*) irradiation are shown

radiation-introduced noise voltage is relatively small compared to the quantization noise in the low conversion rate mode. The current consumption of the TDC system is nearly unaffected during the whole irradiation period. Photos of the substrate and the die before and after 5 MGy total dose irradiation are also shown in Fig. 4.27.

4.7 Conclusions

In this chapter, we have introduced the first $\Delta\Sigma$ TDC which successfully implemented the third-order noise shaping in time domain. It brings us a new type of TDCs, whose resolution is not limited by the intrinsic CMOS gate delay and is not sensitive to the analog component mismatch. The first-demonstrated 1-1-1 MASH $\Delta\Sigma$ TDC is implemented in 0.13 μm CMOS and achieves a time resolution of 5.6 ps, when the *OSR* is 250. It consumes 1.7 mW from a 1.2 V supply and exhibits an *ENOB* of 11 bits. The time resolution of the TDC is mainly limited by the comparator delay, which causes phase skew error when turning on/off the TDC.

Fortunately, this skew error can be calibrated by using an online calibration method. The MASH $\Delta\Sigma$ TDC with delay-line-assisted calibration (TDC-CAL) proposed in this work digitizes the large comparator delay by using the coarse delay line, and makes a real-time compensation to the phase skew. Owing to the power saving in the threshold-detection comparators, the TDC-CAL consumes only 0.7 mW, which is the lowest in the present state of the art and achieves an *ENOB* of 13 bits. A wide input range of 100 ns has also been achieved. Moreover, the resolution of the TDC can be further improved by increasing the *OSR*, without any significant increase in power consumption.

Gamma radiation assessments with both a low dose rate of 1.2 kGy/h and a high dose rate of 30 kGy/h have been performed, proving the TDC's radiation hardness. The system power consumption is almost not affected and even after an extremely high radiation dose of 3.4 MGy, the *ENOB* drops only 1 bit and, for an *OSR* of 250, a 10.5 ps time resolution is still achieved.

Chapter 5
Radiation Hardened Bandgap References

Abstract As a key component in the proposed multistage noise-shaping (MASH) $\Delta\Sigma$ time-to-digital converter (TDC) system, the total ionizing dose (TID) radiation tolerance of the bandgap reference in deep-submicron -complementary metal–oxide–semiconductor (CMOS) technology is generally limited by the radiation-introduced leakage current in diodes. An analysis of this phenomenon is given in this book, and a dynamic base leakage compensation (DBLC) technique is proposed to improve the radiation hardness of a bandgap reference built in a standard 0.13 μm CMOS technology. A temperature coefficient (TC) of 15 ppm/°C from −40 to 125 °C is measured before irradiation. The voltage variation from 0 to 100 °C is only ± 1 mV for an output voltage of 600 mV. Gamma irradiation assessment proves that the bandgap reference is tolerant to a total ionizing dose of at least 4.5 MGy. The output reference voltage exhibits a variation of less than 3 % during the entire experiment, when the chip is irradiated by gamma ray at a dose rate of 27 kGy/h.

5.1 Introduction

Reference voltage generators are critical building blocks in many analog/mixed-signal systems such as A/D, D/A converters, voltage regulators, as well as in the proposed multistage noise-shaping (MASH)$\Delta\Sigma$ time-to-digital converter (TDC). The generators are required to be stabilized over process, supply voltage, and temperature (PVT) variations. For applications in high-energy physics, space, and nuclear reactors, the reference voltage also needs to be stable under high ionizing radiation dose. The complementary metal–oxide–semiconductor (CMOS) bandgap reference technique [44] is one of the most popular solutions which has been successfully implemented to achieve the requirements. However, such a bandgap reference can only provide a fixed voltage of 1.25 V, which limits the low supply voltage (e.g., sub-1-V) operation coming along with the CMOS technology downscaling.

Advanced deep-submicron CMOS technologies have shown an intrinsic enhanced radiation tolerance, due to downscaling of the CMOS gate oxide thickness [47]. However, it has been found that conventional bandgap references in submicron CMOS technology are rather vulnerable to a total ionizing dose (TID) effect [56], due to radiation damage in the diodes.

© Springer International Publishing Switzerland 2015
Y. Cao et al., *Radiation-Tolerant Delta-Sigma Time-to-Digital Converters*,
Analog Circuits and Signal Processing, DOI 10.1007/978-3-319-11842-0_5

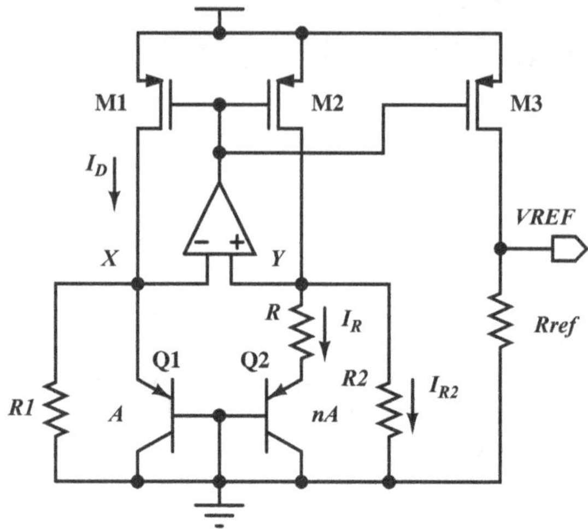

Fig. 5.1 A current-mode CMOS bandgap reference with sub-1-V operation. *CMOS* complementary metal–oxide–semiconductor

In this work, a radiation-hardened bandgap reference with sub-1-V operation is designed in standard 0.13 μm CMOS technology. An analysis on the TID effects in CMOS bandgap references is given. In order to achieve megagray (MGy)-level ionizing radiation tolerance, a dynamic base leakage compensation (DBLC) technique is proposed to eliminate the radiation introduced leakage current in the bandgap core. The remainder of this chapter is organized as follows. An analysis on the TID effects in conventional low-voltage CMOS bandgap references is given in Sect. 5.2. Section 5.3 explains the proposed DBLC technique in detail. Pre-rad measurement and radiation assessment results are discussed in Sect. 5.4. In Sect. 5.5, a conclusion is drawn.

5.2 Total Ionizing Dose Effects in CMOS Bandgap References

5.2.1 CMOS Bandgap Reference with Sub-1-V Operation

A CMOS bandgap reference circuit with sub-1-V operation can be built in a current-mode structure as shown in Fig. 5.1 [8]. A temperature-independent current is first generated by summing the proportional to absolute temperature (PTAT) current I_R and the complementary to absolute temperature (CTAT) current I_{R2}. By applying this current to a reference resistor, a sub-1-V bandgap reference voltage can be obtained. Assuming M1, M2, and M3 have the same size, we note that $I_{\mathrm{D,M1}} = I_{D,M2} = I_{D,M3}$ ($I_{D,Mx}$ is the drain current of transistor Mx), and $V_X = V_Y$ due to the negative feedback. The area of diode Q2 is n times that of Q1. Therefore,

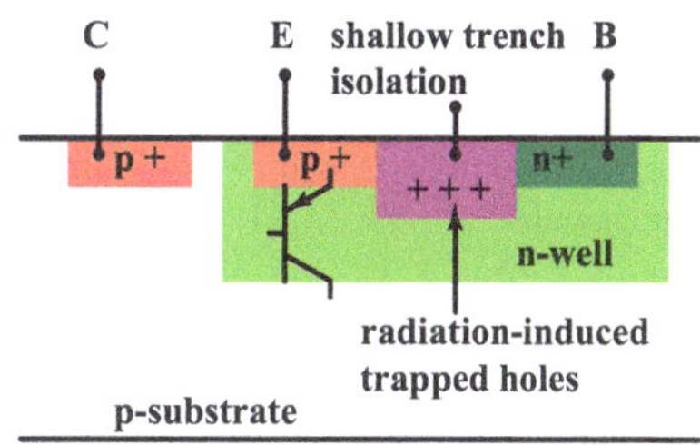

Fig. 5.2 A pnp transistor formed in a n-well CMOS technology

$$
\begin{aligned}
VREF &= R_{ref} \cdot I_{D,M3} \\
&= R_{ref} \cdot \left(\frac{V_Y - V_{EB,Q2}}{R} + \frac{V_Y}{R2} \right) \\
&= R_{ref} \cdot \left(\frac{V_{EB,Q1} - V_{EB,Q2}}{R} + \frac{V_{EB,Q1}}{R2} \right) \\
&= R_{ref} \cdot \left(\frac{V_T \ln \frac{I_{D,M1}}{I_{SS}} - V_T \ln \frac{I_{D,M2}}{n \cdot I_{SS}}}{R} + \frac{V_{EB,Q1}}{R2} \right) \\
&= R_{ref} \cdot \left(\frac{V_T \ln n}{R} + \frac{V_{EB,Q1}}{R2} \right).
\end{aligned} \tag{5.1}
$$

The two diodes are formed by shortening the base and collector of two pnp transistors. In n-well CMOS processes, a pnp transistor can be obtained by doping a p^+ region inside an n-well which serves as the emitter while the n-well itself serves as base, as shown in Fig. 5.2. The p-type substrate acts as the collector and is inevitably connected to the most negative supply.

5.2.2 TID Effects in CMOS Diodes

Although CMOS gate transistors fabricated in deep-submicron technology have shown excellent radiation tolerance, the diode still suffers from radiation-induced leakage current [63]. A shallow trench isolation field oxide layer is usually placed surrounding the p^+ diffusion region, which is the emitter of the pnp transistor. Irradiation-induced holes get trapped in the body of the field oxide near the SiO_2–Si interface. This will increase the base leakage current and degrade the current gain of the bipolar transistor. Consequently, when the bipolar transistors are used in a bandgap voltage reference, the output voltage/current will undergo dramatic changes.

An explanation of this phenomenon can be obtained by analyzing the conventional bandgap reference circuit in Fig. 5.1, represented in Fig. 5.3. When the same small-irradiation-induced leakage current ΔI_B appears at the emitter–base junction of all diodes, the increased emitter currents in the two branches are different, due to the nonidentical numbers of diodes. The increased base leakage current in the left diode is neglected since there is only one unit cell in that branch.

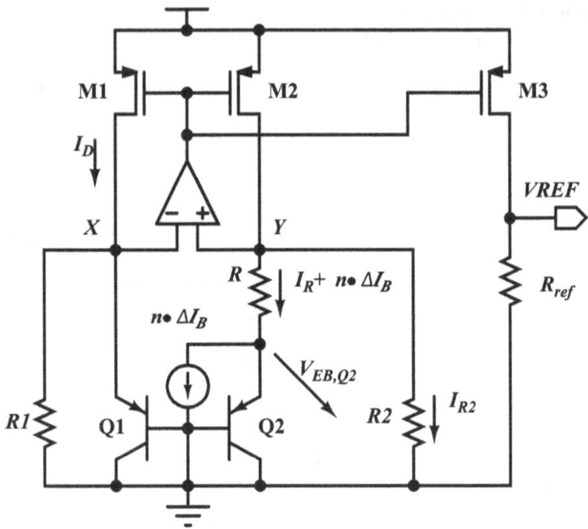

Fig. 5.3 Modeled TID effects in the conventional bandgap reference as diode base leakage currents. *TID* total ionizing dose

Assuming the collector currents I_C remain the same for both diodes after irradiation [63], the emitter–base voltage V_{EB} (the emitter–base voltage of the diode) for both diodes should remain unchanged since $V_{EB} = V_T \cdot \ln(I_C/I_{SS})$. However, the emitter current of the right diode, I_R, which is equal to $(V_{EB,Q1} - V_{EB,Q2})/R$, will affect $V_{EB,Q2}$. The right diode consists of 38 unit cells in this design. Therefore, a significant base leakage current will appear in Q2 due to irradiation. Consequently, I_R has to increase by the same amount to provide this leakage current. This will pull down $V_{EB,Q2}$, when $V_{EB,Q1}$ keeps unchanged. Accordingly, $I_{C,Q2}$ will also decrease until a new balance point is reached. The bandgap voltage:

$$\begin{aligned} VREF &= R_{ref} \cdot (I_R + I_{R2}) \\ &= R_{ref} \cdot \left(I_{C,Q2} + I_{B,Q2} + \frac{V_{EB,Q1}}{R2} \right) \\ &= R_{ref} \cdot \left(I^0_{C,Q2} - \Delta I_{C,Q2} + I^0_{B,Q2} + \Delta I_{B,Q2} \right) \\ &\quad + R_{ref} \cdot \frac{V_{EB,Q1}}{R2}, \end{aligned} \tag{5.2}$$

where $I^0_{C,Q2}$ and $I^0_{B,Q2}$ are the initial collector and base currents of the diode, $\Delta I_{B,Q2}$ is the radiation introduced base leakage current, and $\Delta I_{C,Q2}$ is the collector current reduction due to the $V_{EB,Q2}$ drop.

From Eq. (5.2), one can predict that the reference voltage will increase more rapidly at the beginning of the irradiation, when $\Delta I_{B,Q2}$ is dominant over $\Delta I_{C,Q2}$. As irradiation continues, $V_{EB,Q2}$ drops further, $\Delta I_{C,Q2}$ will become larger. Then it compensates parts of the base leakage current, and smoothes the output reference voltage.

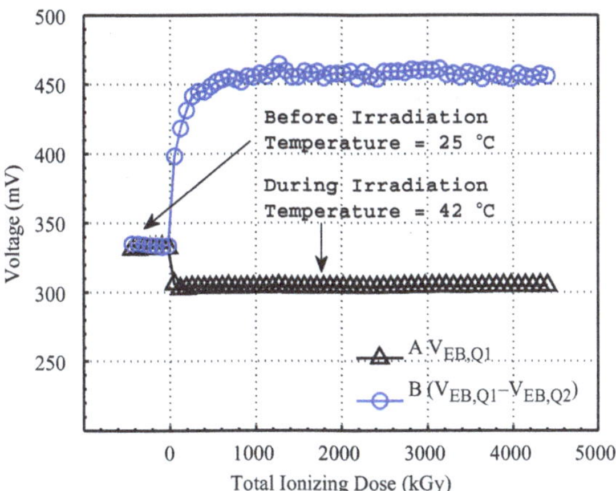

Fig. 5.4 Measured TID effects in the conventional bandgap reference

Figure 5.4 shows the emitter–base voltages of the two diodes in a conventional bandgap reference during a gamma radiation experiment. As derived in Eq. (5.1), by substituting $R_{ref}/R2$ and R_{ref}/R with A and B, the bandgap reference voltage can be rewritten as

$$VREF = A \cdot V_{\mathrm{EB,Q1}} + B \cdot (V_{\mathrm{EB,Q1}} - V_{\mathrm{EB,Q2}}). \tag{5.3}$$

In order to obtain a temperature/radiation-independent bandgap reference voltage, the average value of two components on the right side of Eq. (5.3) has to remain constant. However, in the case of the conventional bandgap reference, a significant drop is found on $V_{EB,Q2}$ while $V_{EB,Q1}$ only has a small shift at the beginning of the irradiation mainly due to a temperature change. Consequently, the output voltage of the bandgap reference increases with the accumulated radiation dose, which is consistent with the theory stated above.

5.3 Radiation-Hardened Bandgap References

5.3.1 DBLC Technique

A previously reported radiation-hardened bandgap reference based on dynamic threshold voltage MOSFET (DTMOS) diodes was proposed in [34] to solve the diodes leakage problem. It replaces the traditional CMOS diodes by DTMOS diodes, which have better radiation hardness. It exhibits good radiation tolerance up to 0.4 MGy. However, for applications in the International Thermonuclear Experimental Reactor (ITER), a total dose radiation tolerance more than 1 MGy is required. The DTMOS diodes are sensitive to high levels of total ionizing dose due to threshold variations affecting diode matching. This is because they are actually P-type metal–oxide–semiconductor (PMOS) transistors operating in the sub-threshold range which

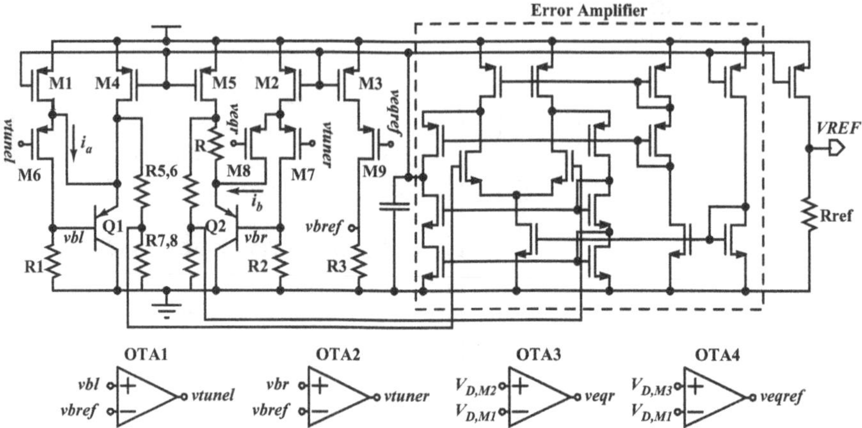

Fig. 5.5 Schematic of a radiation-hardened CMOS bandgap reference using DBLC. *CMOS* complementary metal–oxide–semiconductor, *DBLC* dynamic base leakage compensation

are very susceptible to TID effects [27]. Moreover, DTMOS diodes are not standard devices, which make it more difficult to simulate and evaluate the circuit's performance.

In this work, a dynamic compensation technique is proposed to improve the radiation hardness of the conventional bandgap reference where only normal CMOS diodes are used. The schematic is shown in Fig. 5.5. The purpose of the dynamic compensation unit is to provide all the base current for the diodes. Therefore, only the collector current is flowing in the core circuits. According to early irradiation assessment results [63], the collector current of the diode is nearly unaffected by the increasing ionizing dose. The base leakage currents induced by irradiation will only flow through the compensation circuits, which keeps the bandgap current $I_{D,M4}$ and $I_{D,M5}$ constant.

The concept of using feedback compensation circuits to protect sensitive components in a system is similar to the sensitive node active charge cancellation (SNACC) technique presented in [14]. However, the SNACC technique is useful for single-event mitigation, when the sensitive transistors can be layouted closely in a common-centroid fashion to enable differential charge cancellation [7]. For TID effects mitigation, SNACC is less suitable, since the irradiation-induced leakage current in different devices will only be the same when they have the same characteristic and the same biasing condition. Moreover, for bandgap reference application, a large number of diodes are normally required and they consume a large area. It is impractical to implement differential protective circuitry for each diode as it brings too much area penalty and layouting difficulties. By contrast, the DBLC technique compensates the leakage current produced in the core circuit globally, and it does not need replica devices. This simplifies the design and layout of the core circuitry and can compensate the TID-radiation-induced leakage current in CMOS diodes more accurately.

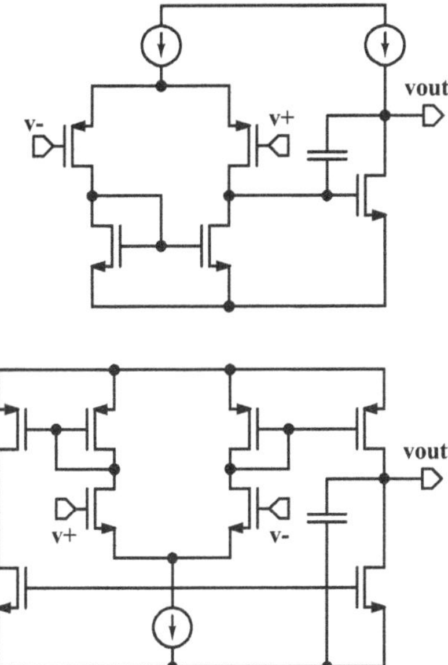

Fig. 5.6 Schematic of the amplifier used in OTA1 and OTA2. *OTA* operational transconductance amplifier

Fig. 5.7 Schematic of the amplifier used in OTA3 and OTA4

5.3.2 Circuit Description

The core of the radiation-hardened bandgap voltage reference remains the same as in the conventional architecture. The feedback error amplifier is implemented in a two-stage folded-cascode structure. It has a gain of 70 dB and a phase margin of 60°, when the power consumption is only 4.8 μW.

The compensation unit works as follows. Resistors R1, R2, and R3 have the same size. Operational transconductance amplifier (OTA)1 and OTA2 are used to keep vbl = vbr = vbref. This is important because $\Delta V_{E,Q1,2}$ is equal to $\Delta V_{EB,Q1,2}$ only when $V_{B,Q1}$ equals $V_{B,Q2}$, where $\Delta V_{E,Q1,2}$ and $\Delta V_{EB,Q1,2}$ are variations of the emitter voltage and the emitter–base voltage of Q1/Q2, respectively. Therefore, $I_{R1} = I_{R2} = I_{R3}$. Meanwhile, M1, M2, and M3 have the same size, which gives $I_{D,M1} = I_{D,M2} = I_{D,M3} = I_{R3}$. We also have $I_{R1} = I_{D,M1} = I_{D,M6} + I_{B,Q1} = I_{D,M6} + i_a$. $I_{B,Q1}$ is thus equal to i_a. For the same principle, $I_{B,Q2}$ also equals i_b. When the base current of the diode increases due to the radiation-induced leakage, it will be compensated dynamically by this base current servo loop. Transistors M8, M9, and amplifiers OTA3, OTA4 are added to further ensure the currents flowing through M1, M2, and M3 are equal.

Schematics of amplifiers OTA1-4 are shown in Figs. 5.6 and 5.7. OTA1 and OTA2 have the same structure, and consume 2.5 μA for each amplifier. They both have a direct current (DC) gain of 60 dB. OTA3 and OTA4 have less required DC gain, which is 35 dB, and each amplifier consumes only 1.2 μA.

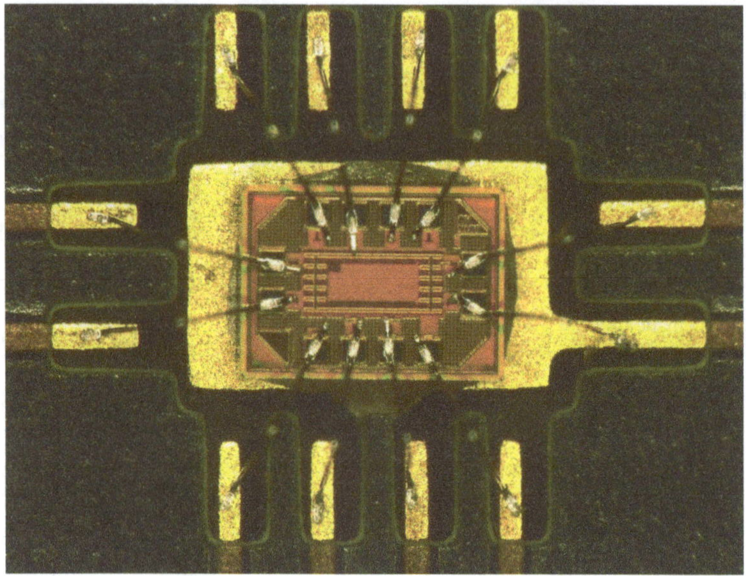

Fig. 5.8 Die photo of the bonded DBLC bandgap reference circuit after irradiation. *DBLC* dynamic base leakage compensation

5.4 Experiment Results

5.4.1 Pre-rad Measurement

The radiation-hardened bandgap reference is implemented in a standard 0.13 μm CMOS technology (the die photo is shown in Fig. 5.8), and it consumes 50 μA (25 μA by the bandgap core, and another 25 μA by the base leakage compensation circuits). The core of the reference circuit occupies an area of only 0.056 mm^2, and the total area is 0.7 mm^2 including all pads. The output voltage of the bandgap reference is designed to be 600 mV. The chip is fully functional within a wide supply voltage range of 0.85–1.5 V, observed in Fig. 5.9.

The measured best TC of the bandgap reference is 15 ppm/°C in a range of − 40 to 125 °C, as shown in Fig. 5.10. From 0 to 100 °C, the voltage variation is only $\pm$ 1 mV. A spread of $\pm$15 mV is found in measurement due to process variation. However, this can be trimmed by varying the on-chip programmable resistor. Figure 5.11 presents the output voltages of the same samples used in the previous measurement after digital trimming. The voltage spread is reduced to within $\pm$4 mV.

5.4.2 Gamma-Irradiation Experiment

More interestingly, the effectiveness of the DBLC technique has also been evaluated by a real-time gamma irradiation assessment. The experiment was carried out

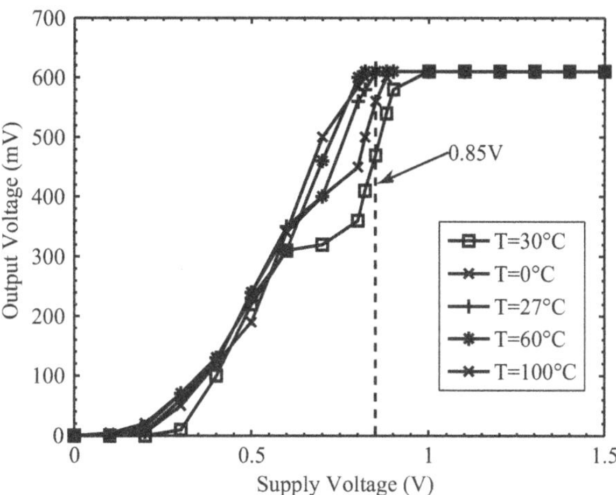

Fig. 5.9 Output voltage versus supply voltage for different temperature

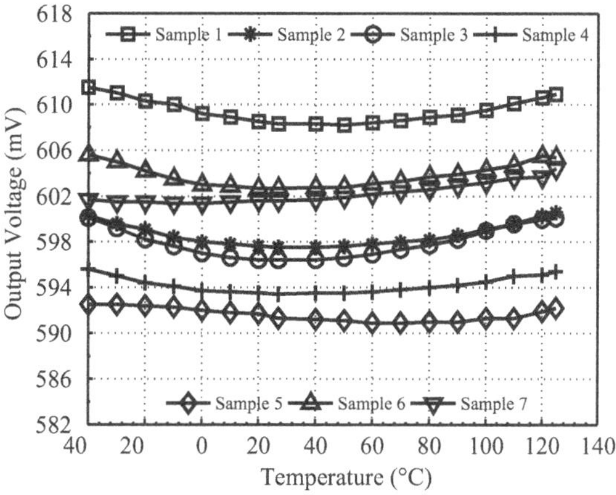

Fig. 5.10 Measured output voltages versus temperature of seven different samples

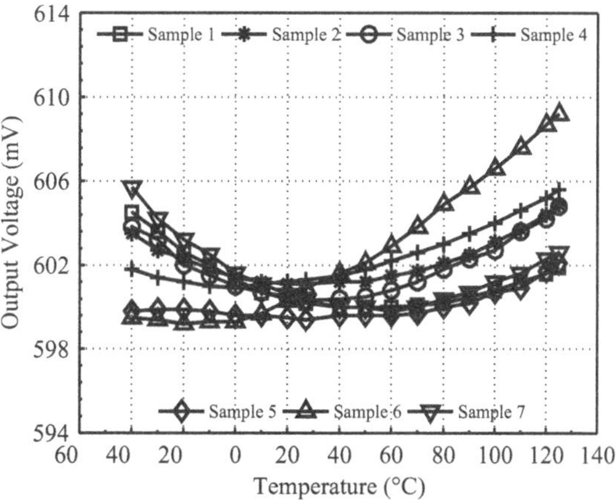

Fig. 5.11 Trimmed output voltages of seven different samples

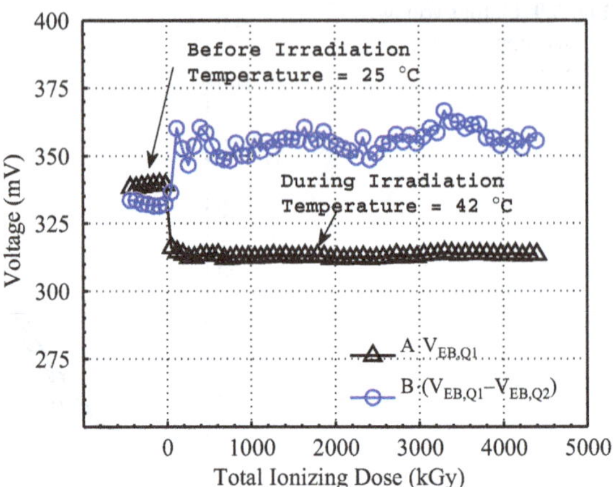

Fig. 5.12 Emitter–base voltages of the diodes in the radiation-hardened bandgap reference using DBLC

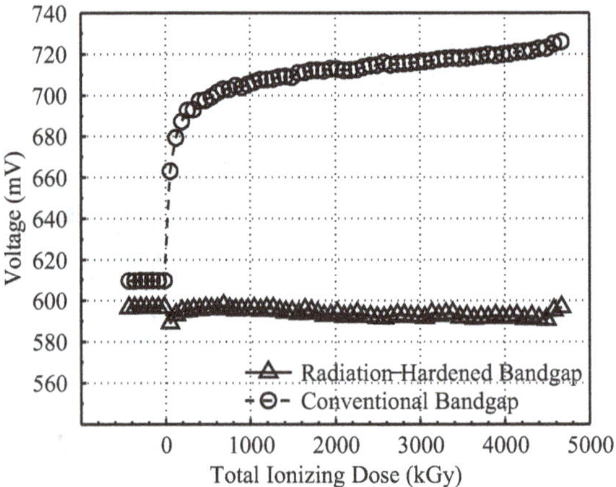

Fig. 5.13 Measured output voltages of the radiation-hardened bandgap reference and the conventional bandgap reference

from the "Brigitte" facility at SCK ·CEN. Substrates bonded with both conventional bandgap references and radiation-hardened bandgap references from the same technology run are irradiated with a ^{60}Co gamma source. A high dose rate of 27 kGy/h is obtained, which enables us to achieve a TID of 4.5 MGy in 1 week.

Figure 5.12 shows the emitter–base voltages of two diodes in the radiation-hardened bandgap reference. Thanks to the DBLC technique, both $V_{EB,Q1}$ and $V_{EB,Q2}$ are immune to the total ionizing dose, where only shifts caused by temperature drift are found. As a result, the output reference voltage of the bandgap employing the DBLC technique stays equal to the pre-rad value. Output voltages of the radiation-hardened bandgap reference and the conventional bandgap reference of one measured sample are shown in Fig. 5.13. A variation of only 1 % is found from 0 to 4.5 MGy. A small dip is observable at the beginning of the irradiation. This is due to the latency existed in the base leakage compensation network which requires a longer time to provide the significantly increased base current, after all diodes are irradiated

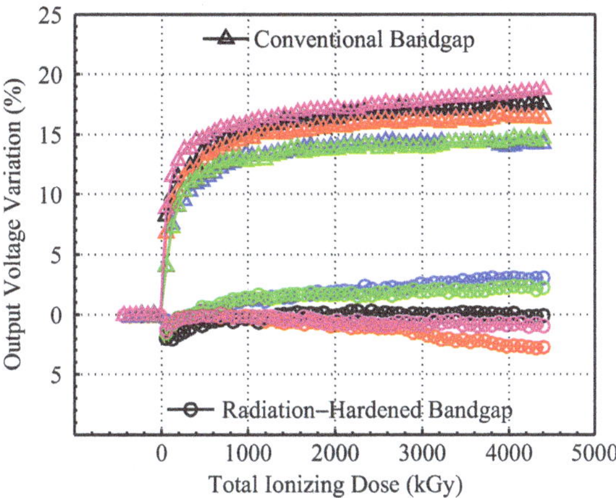

Fig. 5.14 Output voltage variation of five conventional bandgap references and five radiation-hardened bandgap references

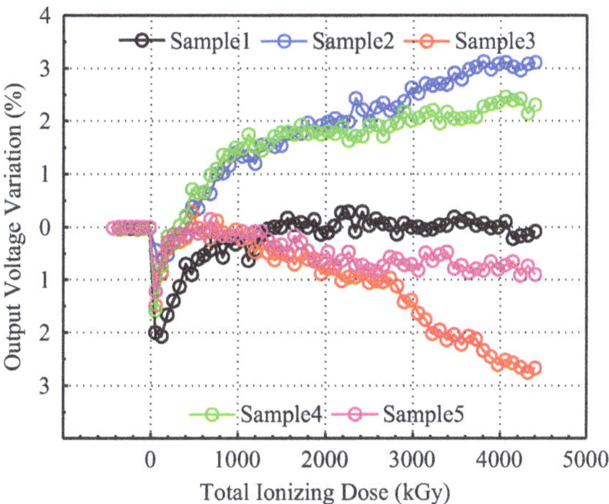

Fig. 5.15 Closed-in inspection of output voltage variation of the DBLC radiation-hardened bandgap references

instantly at such a high-radiation dose rate. On the other hand, the output voltage of the conventional bandgap reference increases more than 15 % during irradiation. After the irradiation stopped, the output voltage of the DBLC bandgap reference can be easily recovered to the pre-rad value by self-annealing.

More than one sample has been measured during the irradiation. The upper group of voltage curves in Fig. 5.14 is obtained from five different conventional bandgap reference samples. They show an average increase of 17 % after irradiation. The lower group of voltage curves which exhibits a variation of less than 3 % is obtained from five DBLC bandgap samples. A closed-in inspection of this variation is shown in Fig. 5.15. Some samples show a slowly increasing slope or decreasing slope in the output reference voltage, which is mainly due to the matching imperfection in the base leakage compensation network.

Table 5.1 Measured specifications

Technology	0.13 μm CMOS
Area	0.056 mm^2 (core)
	0.7 mm^2 (chip)
Supply voltage	0.85–1.5 V
Current consumption	50 μA
Reference voltage	600 mV
Voltage spread	$\pm$ 15 mV
(from seven measured samples)	$\pm$ 4 mV (after trimming)
Temperature range	-40 to 125 °C
Temperature coefficient	15 ppm/°C
Voltage variation caused by	1 % (best case)
γ-irradiation up to 4.5 MGy	$\pm$ 3% (from 5 samples)

CMOS complementary metal–oxide–semiconductor

5.5 Conclusions

In this chapter, a radiation-hardened CMOS bandgap voltage reference is introduced. It provides a reference voltage of 600 mV under room temperature. The bandgap voltage–temperature curve exhibits a temperature coefficient (TC) of 15 ppm/°C from -40 to 125 °C, and the voltage variation from 0 to 100 °C is only $\pm$ 1 mV. The bandgap reference is also compatible with sub-1V operation. A real-time gamma irradiation assessment with a dose rate of 27 kGy/h has been performed to verify the bandgap reference's TID tolerance. The output voltage of the bandgap reference using the DBLC technique has shown a variation of only 3 % after 4.5 MGy.

The measured specifications of the DBLC bandgap reference are summarized in Table 5.1.

Chapter 6
Low-Jitter Relaxation Oscillators

Abstract The best achievable time resolution of the proposed multistage noise-shaping (MASH) $\Delta\Sigma$ time-to-digital converter (TDC) is practically limited by the phase noise performance of the employed relaxation oscillator which generates the quantization clock. After gaining an insight of the noise sources in the relaxation oscillator, a switched-capacitor (SC) integrated error feedback (IEF) technique has been proposed to reduce the closed-in phase noise and improve its clock accuracy. The proposed relaxation oscillator with SC IEF is implemented in 65 nm complementary metal-oxide-semiconductor (CMOS). It demonstrates a phase noise of -101 dBc/Hz at 100 kHz offset frequency, when the center frequency is 12.6 MHz and achieves a figure of merit (FOM) of 152.6 dB. The oscillator occupies an area of 0.01 mm^2, and consumes 82 μA from 1.2 V. The frequency changes only $\pm$0.07% over a supply range of 1.1–1.5 V. From 0 to 80 °C, the total frequency variation is within $\pm$0.82 %.

6.1 Introduction

As discussed in Chap. 4, the best achievable time resolution of the proposed multi-stage noise-shaping (MASH) $\Delta\Sigma$ time-to-digital converter (TDC) is highly relying on the relaxation oscillator's phase noise performance. Two major factors which effectively limit the TDC's signal-to-noise ratio (SNR) are the timing jitter and phase skew both present in the relaxation oscillator. The relation between the oscillator's phase noise and the TDC's SNR has been explained in Sect. 4.3. In order to further improve the TDC's resolution, the relaxation oscillator's phase noise has to be investigated and optimized.

Besides its application in the MASH TDC, the relaxation oscillator has also been used as voltage-controlled oscillator (VCO) in a phase-locked loop (PLL) and an on-chip reference clock generator. The growing demand in implementing fully integrated low-cost low-power area-efficient system on chips (SoCs), such as wireless sensor node (WSNs), implanted biomedical devices, and microcomputers, has again brought many interests in searching for on-chip reference clock generation solutions. Relaxation oscillators are suitable candidates to generate such reference clocks due to their compact size, low power consumption, and wide frequency tuning range. However, the poor phase noise performance and large long-term variation are two major problems which limit their application.

© Springer International Publishing Switzerland 2015

Y. Cao et al., *Radiation-Tolerant Delta-Sigma Time-to-Digital Converters*,
Analog Circuits and Signal Processing, DOI 10.1007/978-3-319-11842-0_6

In this chapter, an anti-jitter technique based on the switched-capacitor (SC) integrated error feedback (IEF) is introduced to improve the relaxation oscillator's phase noise performance. The relaxation oscillator employing the SC IEF architecture is suitable for both the MASH TDC and on-chip reference clock generation applications. In order to gain a solid understanding of relaxation oscillators, some background knowledge with regard to their applications as on-chip clock references and performance measures is also given in the chapter.

6.2 On-Chip Clock Generation

Nowadays, a SoC contains almost all individual blocks including analog, digital, and radio frequency (RF)—which are essential to achieve its full functionality—on a single silicon die. However, there are still some blocks which are not suitable for a single-chip integration. Among them, a good example is the reference clock. Crystal oscillators (XOs) have been serving as the primary reference clock of all electronic systems for a long history. Their ultimate phase noise performance (< −130 dBc/Hza rate of 1 kHz) and excellent temperature stability (< ± 50 ppm@ −40 to 125 °C) have put them on an unbeatable position in the clock references market.

But for certain applications, such as in biotelemetry systems and WSN, a reference clock with such a high specification is often not required [23, 93]. On the other hand, on-chip fully integrated solutions are preferred due to the strict power, area, and cost constraints. Many efforts have been put in investigating alternative reference clock generation solutions in recent years, and several promising techniques were proposed to be implemented in future fully integrated SOCs. The three most interesting ones are reviewed here, which are microelectromechanical system (*MEMS*) oscillators, inductance–capacitance (*LC*) oscillators, and ring oscillators.

6.2.0.1 MEMS Oscillators

Recent advancement in microelectromechanical systems enables producing high-quality silicon resonators. A *MEMS* resonator operates in a similar principle to the mechanical tuning fork used to tune musical instruments. A separate electronic oscillator provides impetus that forces the *MEMS* resonator to vibrate at a precise frequency. A PLL captures and distributes the reference clock generated by the resonator–oscillator combination.

A key issue of the *MEMS* resonator frequency is that it varies with temperature. In [64], this temperature variation is suggested to be corrected by using a temperature-to-digital converter. It demonstrates a better than ± 0.5 ppm frequency accuracy over the industrial temperature range, which is very close to the performance of high-end quartz-based clock references. However, this outstanding performance comes at the cost of high power consumption (100 mW for a 48 MHz output clock) and large area (3.8 mm^2). For low-power area-efficient SoCs, this is generally unacceptable.

6.2.0.2 LC Oscillators

Another attracting approach to implement a monolithic XO-class on-chip clock generator is the self-referenced radio frequency *LC* oscillator [54]. It utilizes the self-oscillation frequency of an *LC* tank circuit to provide the reference signal, and tackles its frequency drifting problem due to variations in temperature and bias conditions by applying compensation techniques.

The proposed *LC* oscillator achieves an approximately ± 400 ppm total frequency accuracy from -10 to $+85$ °C, and a good phase noise performance of -103 dBc/Hz at 10 kHz. The specifications make it suitable for applications with moderate clock accuracy requirements, such as Universal Serial Bus (USB) 2.0, peripheral component interconnect (PCI), etc. Unfortunately, the power consumption of this *LC* clock generator is still considerably high (31 mW) for most low-power SoC applications.

6.2.0.3 Ring Oscillators

A well-known low-cost area-efficient on-chip clock generation solution is the complementary metal-oxide-semiconductor (CMOS) ring oscillator. For a long time, the poor process, supply, and temperature variation susceptibility has been the major obstacle to prevent its widespread usage as a clock reference. A compensation technique described in [84] has improved the clock accuracy of a ring oscillator to an order of ± 2.6 % across a temperature range of -40 to 125 °C. However, the poor clock stability, mainly the large close-in phase noise, hampers it being used as a primary clock reference source even for low-standard wireless systems such as biotelemetry and WSN. Moreover, to generate a 7-MHz clock, the system consumes more than 1.5 mW, which is still too high for low-power applications.

6.3 Performance Measures on Clock References

6.3.1 Clock Stability and Accuracy

Two key performance measures of a reference clock are clock stability and clock accuracy. The clock stability describes how well the oscillator frequency resists fluctuations. The dominant factors that affect stability are circuit noises, random variations in temperature, and supply voltage. The clock accuracy describes how well the actual frequency matches the specified frequency under different supply, temperature, and process settings. Figure 6.1 illustrates the effects of clock stability and accuracy in time domain.

In each of the diagrams, the dotted line represents the specified, desired frequency, and the solid line represents the actual frequency generated by the oscillator. The upper-left diagram shows an inaccurate and unstable clock; the actual frequency is not centered around the desired frequency, and it changes with respect to time. The

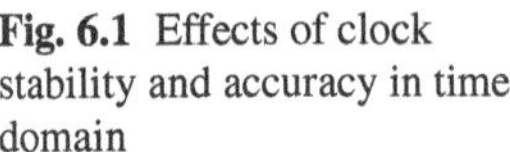

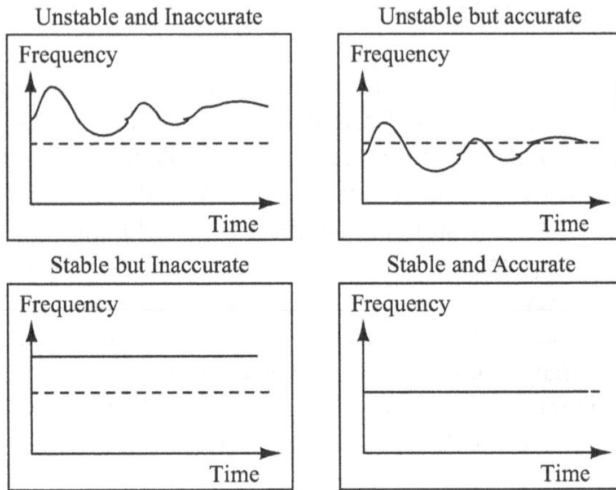

Fig. 6.1 Effects of clock stability and accuracy in time domain

upper-right diagram shows an accurate but unstable clock; the actual frequency is centered around the specified frequency but still changes with time. Conversely, the lower-left diagram shows a clock that is inaccurate but stable; the actual frequency is not centered about the desired frequency, but the output frequency from the oscillator does not change over time. A perfect oscillator would generate a frequency like that presented in the lower-right diagram, where the clock is accurate about the desired frequency and does not change with time.

6.3.2 Phase Noise and Jitter

The random frequency fluctuation of a clock reference can also be characterized as phase noise. For an ideal oscillator operating at ϖ_0, the spectrum has the shape of an impulse. In reality, the spectrum always exhibits "skirts" around the center frequency. This is due to the presence of various noise sources in the circuitry. In particular, low-frequency noise sources, such as $1/f$ noise, will be upconverted into close-in phase noise. The wide-band *white* noise, shaped by the transfer function of the oscillator system, makes up the noise floor at large offset frequencies [67]. In CMOS oscillators, $1/f$ noise is expected to have a significant impact on the overall phase noise performance due to the large flicker noise exhibited in CMOS transistors. Therefore, in order to improve the clock stability, low-frequency noises in the oscillator circuit has to be eliminated or minimized.

The jitter is the time domain counterpart of the phase noise specification. Jitter can be expressed in many ways, such as period jitter, cycle-to-cycle jitter, and long-term jitter. Among them, period jitter is the time difference between a measured cycle period and the ideal cycle period. Due to its random nature, this jitter is often measured by the root of mean square (RMS). In circuit design literature, the period jitter of an oscillator is often specified by the normalized one-period time error (often expressed in parts per million):

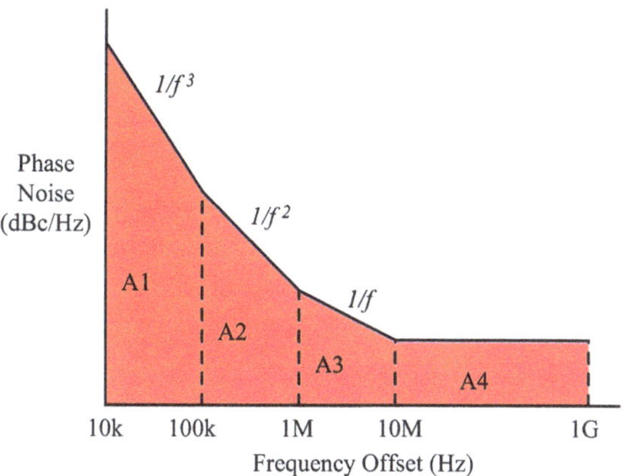

Fig. 6.2 Illustration of phase noise

$$Jitter = \frac{\sigma_{\Delta T_{OSC}}}{T_{OSC}}. \tag{6.1}$$

The RMS period jitter can be converted from the oscillator phase noise. A typical phase noise graph is shown in Fig. 6.2. By integrating the phase noise power (dBc) over the frequency range of interest, i.e., $A = 10\log(A1 + A2 + A3 + A4)$, the RMS jitter can be calculated as [24]:

$$\text{RMS period jitter (seconds)} = \frac{\sqrt{2\times 10^{A/10}}}{\varpi_0}. \tag{6.2}$$

The long-term jitter, which is usually referred to as the accumulated jitter, measures the change in a clock's output from the ideal position, over several consecutive cycles. It differs from period jitter because it represents the cumulative effect of jitter on a continuous stream of clock cycles over a long-time interval. The growth of the accumulated jitter strongly depends on the nature of the noise, and especially on its autocorrelation function. This autocorrelation function indicates how much correlation exists between a noise signal and a time-shifted version of the same noise signal. If no correlation exists, the total accumulated jitter variance is equal to the sum of the variances, $\sigma_{acc}^2 = N\times\sigma^2$. However, if timing errors are correlated, the accumulated jitter increases linearly with the number of consecutive periods, as shown in Fig. 6.3. Hence, depending on an observed long-term jitter over certain number of periods, whether correlated or uncorrelated noise contribution dominates the accumulated jitter can be determined. For example, if a large $1/f$ noise presents, the measured accumulated jitter should be dominated by the correlated noise.

6.4 An Short Review of Relaxation Oscillators

6.4.1 Relaxation Oscillators as VCOs

Relaxation oscillators have a long history of being used as VCOs in PLL circuits for telecommunication applications, such as frequency modulation and demodulation.

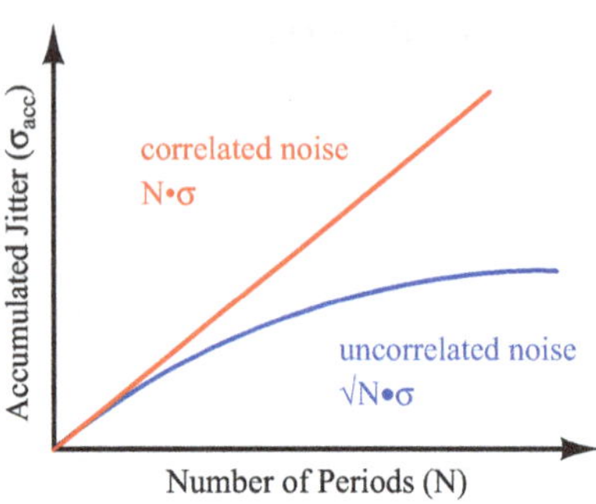

Fig. 6.3 Accumulated jitter dominated by correlated or uncorrelated noise

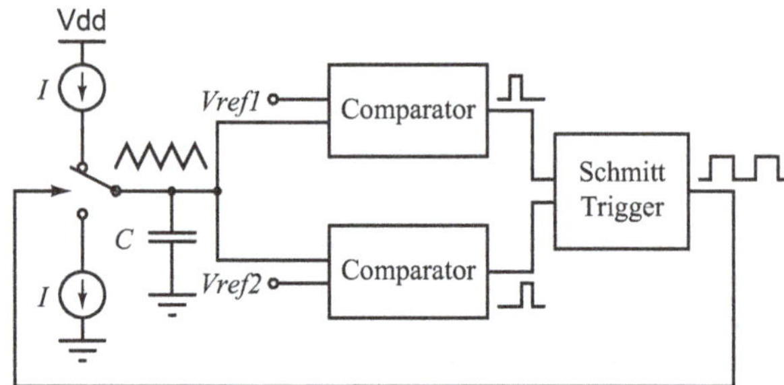

Fig. 6.4 Low-noise grounded capacitor relaxation oscillator

The key advantages of relaxation oscillators are a high control linearity and a large frequency tuning range. A simple low-noise grounded capacitor relaxation oscillator topology is shown in Fig. 6.4 [3]. The frequency of the generated clock can be simplified as:

$$f = \frac{2I/C}{V_{ref1} - V_{ref2}}, \tag{6.3}$$

where *I* is the charging current, *C* is the so-called timing capacitor, and *Vref1,2* are the reference voltages. Following a similar noise analysis of the relaxation oscillator given in Chap. 4, the jitter is determined by the noise voltage associated with the input of the comparator, and the noise current in parallel with the current source I. The frequency of the reference clock can be controlled by either tuning *Vref1,2* or *I*. Theoretically, the oscillator could achieve a control linearity as good as the tuning voltage or current.

However, Eq. (6.3) did not reveal all components in the complete frequency expression of the relaxation oscillator. In reality, the delay of the comparator and the Schmitt trigger also contribute to the total period of the generated clock. This delay is process and signal dependent and does not change linearly with the tuning voltage or current, thus deteriorating the oscillator's control linearity. In order to tackle this problem, different feedback mechanisms have been proposed in [2, 75, 92]. The similarity between these approaches is that they all convert the reference clock frequency back into a voltage, and compare it with another reference voltage provided by an external voltage source or an on-chip bandgap reference. The error

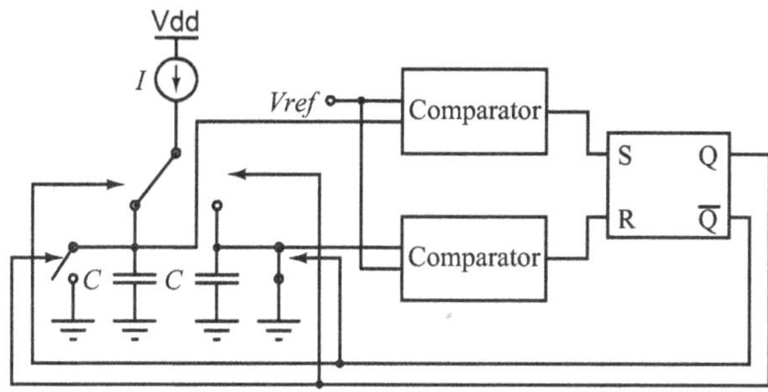

Fig. 6.5 Double-capacitor relaxation oscillator

information between those two voltages is then added to an auxiliary tuning port as feedback. This method can indeed improve the oscillator's control linearity by a large margin. However, it is less effective in improving the oscillator's jitter performance. This is because extra error has been introduced during the frequency to voltage conversion process, and more significantly, the added active feedback circuitry also contributes noise into the oscillator system which cannot be removed out by itself.

In [28], the one grounded capacitor relaxation oscillator has been extended into a double-capacitor structure, as shown in Fig. 6.5. The latter offers several new benefits: (1) only one reference voltage is required instead of two; (2) the amplitude of the capacitor voltage is no longer limited by the two threshold voltages; (3) the desired 50 % duty cycle is easier to achieve after removing the P-type metal-oxide-semiconductor (PMOS) charging and N-type metal-oxide-semiconductor (NMOS) discharging current mismatch problem. All those advantages make the double-capacitor structure a more favorable choice for low-voltage applications.

Another type of relaxation oscillator, the coupled sawtooth oscillator, is presented in [32] that combines excellent control linearity with low timing jitter. Conventionally, high control linearity is achieved by making the threshold-detection comparator fast, resulting in a small delay t_d compared to the overall period of oscillation. In this work, a new level detection concept is introduced: instead of aiming to instantly reverse the capacitor current upon a threshold level crossing, a differential pair is used to gradually turn on the capacitor's charge current. This approach allows low jitter to be achieved without deterioration of the relaxation oscillator's intrinsically high control linearity. However, this control mechanism is not suitable for our application. The relaxation oscillator is also considered to be implemented in the MASH TDC system besides as a reference clock, where instantaneously turning on/off the capacitor's charging current is necessary.

6.4.2 Clock Generation Using Relaxation Oscillators

In recent years, the usage of relaxation oscillators as on-chip reference clocks has been thoroughly investigated, mainly attracted by their compact size, low power consumption, and wide frequency tuning range. In applications such as implantable biotelemetry and ultra-low-power (ULP) radios, where SOCs are designed under very stringent power and area constraints, a fully integrated low-power clock source is highly desired.

However, compared to an XO, the poor phase noise performance and high temperature/supply sensitivity of a conventional relaxation oscillator impose significant challenges on their implementation as system primary clock source. Several attempts have been made over the past decade to improve the stability and accuracy of relaxation oscillators. In [10], a nano-power relaxation oscillator is designed for ultra-high frequency radio-frequency identification (UHF RFID) transponder application. A low-voltage inverted mirror feedback V_{GS}/R reference is proposed to provide correlated charging current and voltage references for the oscillator. As a result, the oscillator frequency is solely determined by the resistor in the reference and the timing capacitor to meet the frequency tolerance specification. However, several drawbacks limit its effectiveness: First of all, this approach does not cover the delay of the comparator, which must be considered as part of the clock period when a megahertz clock is generated. Second, the temperature stability of the oscillator is not addressed. Third, the phase noise performance is not improved.

Similarly, [73] suggests another interesting approach to generate a very accurate reference clock for crystal-less ULP radios, by employing a mobility-based current reference to provide the charging current I. After a single-point calibration, the spread of its output frequency is less than 1.1 %. Unfortunately, the phase noise performance is not discussed in this work, and the delay variation of the comparator is also not addressed. In [21], a precision relaxation oscillator with a self-clocked offset-cancellation scheme for implantable biomedical SOCs is proposed. An auto-zero technique is used to tackle the offset and noise associated with the comparator, resulting in smaller frequency drifts and lower close-in phase fluctuations. But the phase noise suppression is restricted to a limited bandwidth, and it only shows a maximum 8 dB improvement from 1 to 20 kHz offset frequency. Furthermore, its temperature and supply stability can only be enhanced through manual tuning of individual samples, which is impractical and implies a high cost.

As a conclusion, in order to make a relaxation oscillator a practical and reliable reference clock, its phase noise performance and frequency stability have to be considered at the same time. In the next section, low-jitter relaxation oscillators with a closed-loop feedback mechanism are investigated, which has been proved as an effective way to enhance both the stability and accuracy of the clock.

6.4.3 Low-Jitter Oscillator Design

In [30], a SC filtering technique is suggested to filter the jitter noise in a conventional single-grounded-capacitor relaxation oscillator. It uses a similar principle of an anti-jitter circuit technique proposed in [89], which is illustrated in Fig. 6.6. The integrator converts the input clock signal into a sawtooth waveform (Fig. 6.6c) with a constant

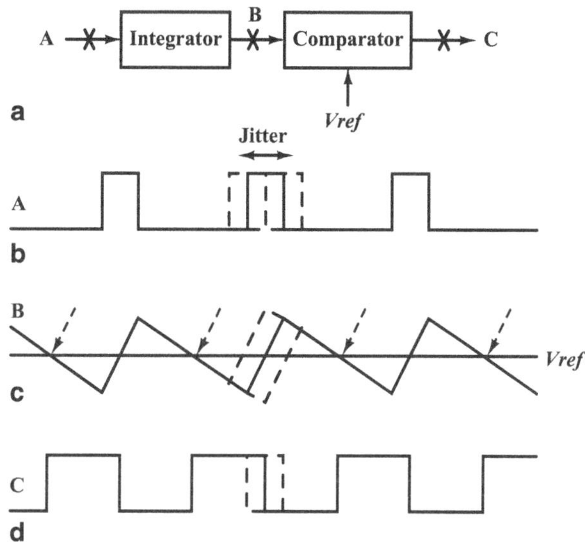

Fig. 6.6 Anti-jitter technique principle: **a** block diagram, **b** input clock waveform at point A, **c** integrator output at point B, and **d** filtered output waveform after the comparator

mean direct current (DC) level, *Vref*. The comparator is arranged to switch at Vref. Figure 6.6d shows that timing jitter (as shown in Fig. 6.6b) on a pulse does not affect the comparator switching time of the downslope ramp section of the sawtooth waveform. Thus, the output clock generated by the comparator triggered at the time of this intersection will have a significantly reduced time jitter. In [30], this action is done by subtracting a fixed charge packet from the charging capacitor. However, this approach still requires a low-noise high-speed comparator to generate the final output clock, which can be power hungry and will practically limit the oscillator's overall phase noise performance.

The power averaging feedback (PAF) technique presented in [87, 88] provides another solution to reduce the oscillator's low-offset phase noise. In the PAF configuration, an active resistor–capacitor low-pass filter (RC LPF) is employed to convert the oscillation frequency into a voltage, whose slow time-varying amplitude represents the close-in phase fluctuations. Large RC values are thus required to enable accurate extraction of the waveform's DC component, which consumes more area and moreover, its response time to high-frequency noise is rather long due to the limited bandwidth.

In this work, we introduce an SC IEF technique to overcome the shortages in aforementioned methods. It employs an SC feedback loop to suppress the offset and flicker noise in the threshold-detecting comparator, resulting in an outstanding phase noise performance at low-offset frequency in its belonging category. Meanwhile, the power and area consumption are minimized to make it applicable for low-cost area-efficient SoC applications. The proposed structure is also fully compatible with the MASH $\Delta\Sigma$ TDC's operating mechanism.

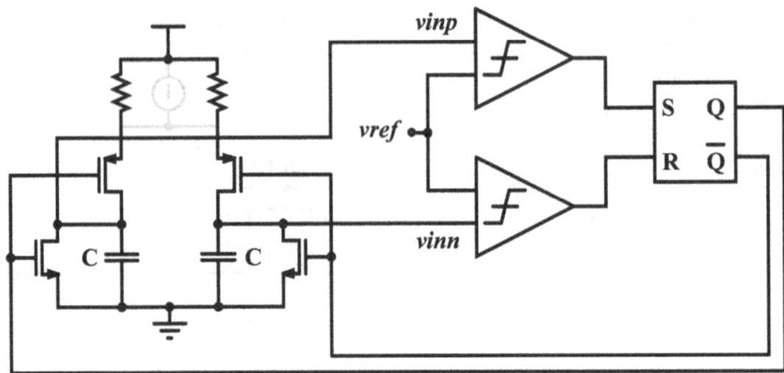

Fig. 6.7 Phase noise optimization step 1: replacing the current source with resistors

6.5 The Relaxation Oscillator with SC Integrated Error Feedback

6.5.1 *Phase Noise Optimization*

Identification of all the noise sources and critical components in a conventional relaxation oscillator will allow us to find out the best solution to solve aforementioned issues. As discussed in Chap. 4/Sect. 4.3, two primary noise contributors to the relaxation oscillator's jitter—hence phase noise—are the noise current in parallel with the charging current source, and the noise voltage in series with the threshold level. The process of optimizing the oscillator's phase noise will follow the principle of mitigating these noise sources or limiting their impact on the oscillator's performance.

Figure 6.7 shows a conventional double-grounded-capacitor relaxation oscillator circuit with the charging current source (drawn in gray) omitted. In general, the current source causes several issues: (1) the flicker noise presented in the active current source deteriorates the close-in phase noise performance; (2) aging of the current source will degrade the accuracy of the generated clock; (3) extra calibration circuitries are required in order to provide a stable charging current over process–voltage–temperature (PVT) variations.

Alternatively, the currents which charge the capacitors can be directly derived from a resistor connected to the supply, which eliminates the aging effect and noise associated with an active current source [87]. Neglecting the delay of the comparator t_d, the period of the oscillation clock can be written as

$$T_{osc} = 2\times RC\times ln(1-\alpha)^{-1}, \tag{6.4}$$

where α equals *Vref/Vdd*. Equation (6.4) means that the oscillation frequency does not have any sensitivity to *Vdd* if α is fixed. This can be done by generating *Vref* through a resistive divider connected between *Vdd* and ground. With temperature-independent capacitors (e.g., metal-oxide–metal capacitors in standard

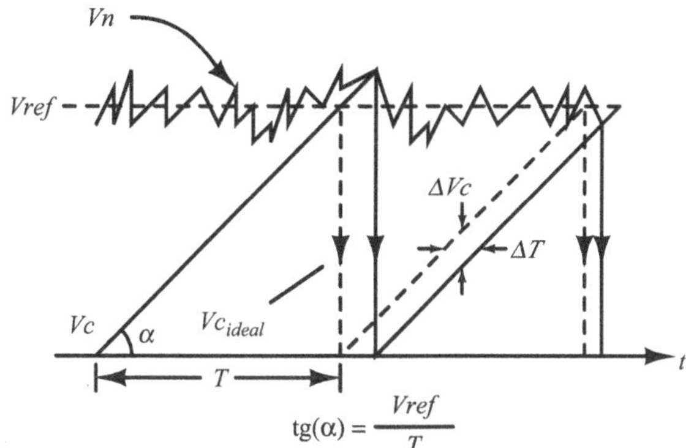

Fig. 6.8 Conversion of the threshold level noise into timing error

CMOS processes), Eq. (6.4) implies that the oscillator's temperature sensitivity is only dominated by *R*. Hence, good temperature stability can be achieved by carefully choosing a resistor device with low sensitivity or applying an appropriate temperature-tuning gradient to the resistor.

After removing the active current source from the conventional oscillator circuit, the comparator now becomes the primary contributor to the oscillator's phase noise due to its large noise bandwidth [31]. The noise in the comparator can be modeled as a noise voltage in series with the threshold reference level. First, let us examine the case in a conventional relaxation oscillator shown in Fig. 6.5. When the capacitor voltage V_c is in the vicinity of the reference voltage, the comparator amplifies the voltage difference between these two signals, and eventually reverts the output status of the trigger circuit (e.g., a S-R latch). This is the so-called oscillator *regeneration*. A general approach to calculate the jitter of such an oscillator is summarized as the *first crossing approximation* [3]. It suggests that the first crossing of the reference voltage (plus noise) by the capacitor voltage invokes an immediate and infinitely fast regeneration. It is assumed that the process of the regeneration itself introduces no additional jitter, but only a constant delay.

Figure 6.8 shows the conversion of the comparator threshold level noise into timing error. The noise voltage V_n, which is present on the reference voltage V_{ref} in Fig. 6.5, affects the timing of the capacitor voltage waveform. When the capacitor voltage crosses the reference voltage plus noise, the oscillator regeneration begins. This will start a new ramp on the second charging capacitor, exhibiting a time error ΔT to the ideal waveform V_{Cideal}. The momentary value of the noise voltage V_n at the crossing moment translates into an equivalent error in the capacitor voltage ΔV_C. The capacitor's charging slope determines the ratio between the capacitor voltage error ΔV_C and the time error ΔT according to:

$$\mathrm{tg}(\alpha) = \frac{dV_C}{dt} = \frac{\Delta V_C}{\Delta T}. \tag{6.5}$$

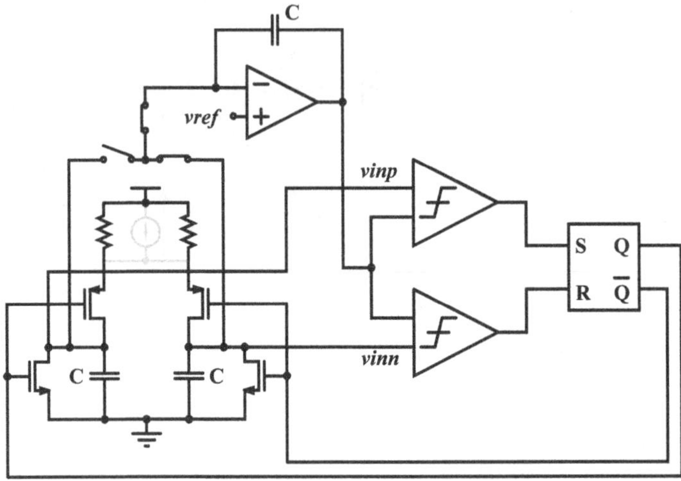

Fig. 6.9 Phase noise optimization step 2: implementing the integrated error feedback

Since the capacitor's charging slope can also be expressed as

$$\frac{dV_C}{dt} = \frac{V_{ref}}{T}, \tag{6.6}$$

the resulting time error is thus equal to

$$\Delta T = \frac{\Delta V_C \cdot T}{V_{ref}}. \tag{6.7}$$

This indicates that, in order to minimize the time error, the reference voltage should be kept as high as possible within the constraints of the circuit. When the charging current source in the conventional oscillator is replaced by resistors as suggested earlier, the same principle still applies. However, in modern CMOS technologies, the increase of the reference voltage is practically limited by the low supply voltage and the small input common-mode range of the comparator's differential pair. Thus, the only way to reduce the comparator's noise is to apply huge power (for thermal noise) and large size (for flicker noise) to its input gain stage, which are not appreciated by low-power area-efficient SoC applications.

Fortunately, these issues can be solved by employing an IEF technique, as shown in Fig. 6.9. The IEF is realized by a SC circuit, and it acts as a LPF to extract the low-frequency phase error information. It is inserted between the capacitor voltage node and the comparator threshold reference node. The IEF is configured in a way such that the dynamically generated reference level V_c can compensate the momentary noise voltage exhibited at the input node of the comparator. Thanks to its negative feedback effect, the phase noise spectrum of the oscillator in the low-offset frequency band is suppressed by the high-pass filter-like transfer function of the IEF. Therefore, we can keep the comparators small and low power. The close-in phase noise is now

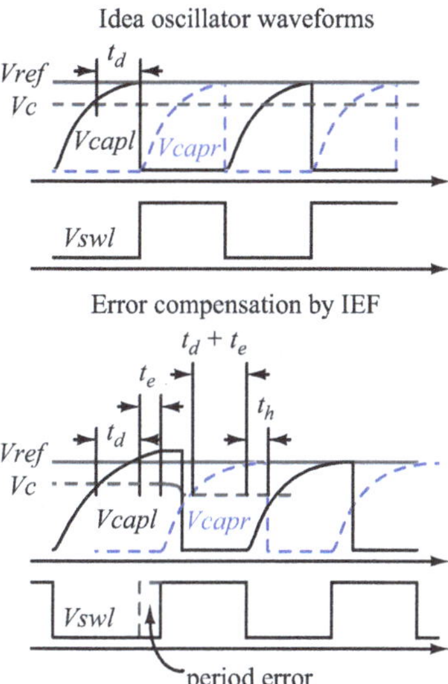

Fig. 6.10 Illustration of the error compensation mechanism

dominated by the low-frequency noise of the feedback amplifier in the SC filter, which can be suppressed through chopper stabilization or auto-zeroing. Comparered to the open-loop approach proposed in [21], this closed-loop configuration provides better jitter suppression and enhances the oscillator's clock accuracy.

The detailed working principle of the IEF is explained in Fig. 6.10. Assuming a low-frequency noise voltage appears at the reference node of the comparator, it causes a small delay variation t_e in addition to the fixed comparator delay t_d. Consequently a time error, equal to t_e, is present in the oscillator clock period. The IEF compares the final voltage at the capacitor Vcapl or Vcapr with the reference voltage *Vref* and integrates their difference ve by using the SC circuits. In a short time, the charging slope on the capacitor can be considered constant when its voltage level approaches *Vref*, which is set close to *Vdd*. Therefore, ve is linearly related to t_e. By subtracting ve from Vc, which is the reference level at the comparator threshold node, the clock period error can be compensated. Not only is the IEF effective at suppressing the low-frequency noise but also improves the oscillator's stability against voltage, temperature variation, and radiation due to the closed-loop structure.

Another important modification to the conventional architecture is using only one threshold-detecting comparator instead of two to detect the switching threshold, as shown in Fig. 6.11. This gives an advantage to enable effective suppression of low-frequency noise (e.g., flicker noise) by using error feedback. Since in the two comparators case (one comparator for each left and right branch) the noise generated in each device is uncorrelated and does not cancel out, it makes the compensation to the clock period error inaccurate.

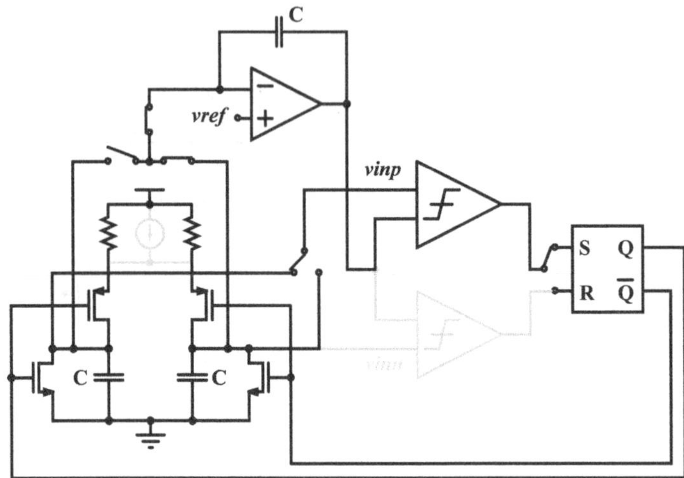

Fig. 6.11 Phase noise optimization step 3: using single threshold-detecting comparator

6.5.2 System Implementation

The complete schematic of the proposed relaxation oscillator with SC IEF is shown in Fig. 6.12. The core of the relaxation oscillator remains the same as in a conventional two-grounded-capacitor structure. Instead of using an active current source, the charging current is directly derived from the power supply through two big resistors. The size of the charging resistors and timing capacitors are determined by Eq. (6.4) for a given clock frequency specification. Further explanation of how to choose appropriate *R* and *C* values will be given shortly in this section.

The SC integrator first samples the momentary value of the timing capacitor voltage at the clock transition moment by enabling one of two switches connected to Vfb. This voltage is then compared with the reference level *Vref*. The difference between these two voltages is integrated at the output of the feedback amplifier to generate the threshold voltage Vc for the main comparator cmp1. The error integrating operation is performed in a predefined integrate-and-hold (IH) phase t_h. This is realized by using only few additional control units. The main purpose of the control logic is to guarantee one capacitor is reset to "0" before the other reaches *Vref*. Then, the alternate charging process can be sustained without being disrupted. This is secured by using a second comparator cmp2 which has been set to a lower threshold Ven to monitor the voltages at the capacitors. Once it crosses Ven, the comparator generates a signal to terminate the integrating phase and turns the SC integrator into hold. Since cmp2 is only used to produce clock signals for the IH circuits, its noise specification has no influence on the oscillator's overall phase noise. The basic operating waveforms of the control logic are illustrated in Fig. 6.13.

Looking at Eq. (6.4), the oscillation period, T_{osc}, is determined by the value of *R*, *C*, and α, which represent the charging resistor, the timing capacitor, and the reference

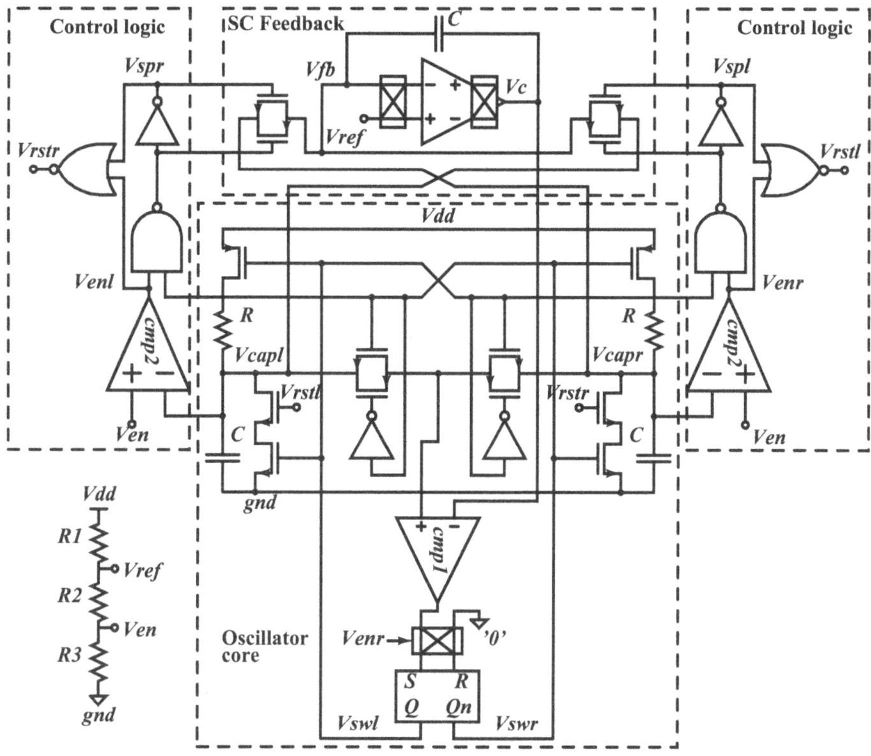

Fig. 6.12 Schematic of the relaxation oscillator with switched-capacitor integrated error feedback

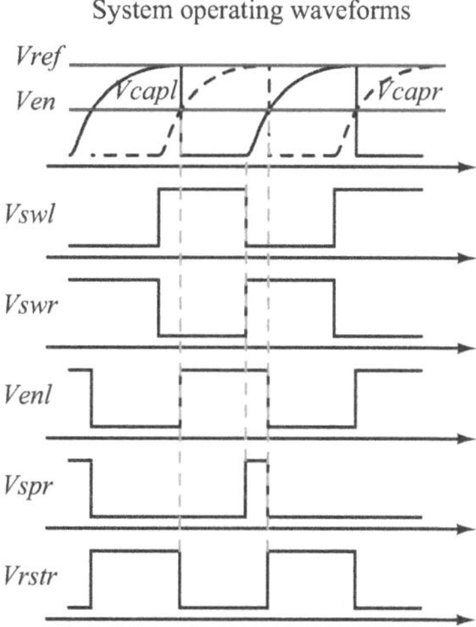

Fig. 6.13 Operating waveforms of the relaxation oscillator with switched-capacitor integrated error feedback (*SC IEF*)

to supply voltage ratio, respectively. As explained earlier, a higher reference voltage will result in smaller jitter. Consequently, we should maximize α for a given specified oscillation period within the constraints of the supply voltage. Accordingly, the value of R and C need to be kept as low as possible. However, when the power consumption is taken into account, a different solution is found. The transient voltage on the timing capacitor, Vcap, can be written as

$$V_{capl,r}(t) = Vdd(1 - e^{-t/RC}). \tag{6.8}$$

The transient charging current is equal to

$$I(t) = \frac{Vdd \times e^{-t/RC}}{R}. \tag{6.9}$$

Thus, the rms value of the oscillator's current consumption in one-half period $T = T_{osc}/2$ can be expressed as

$$\begin{aligned} I_{rms} &= \frac{1}{T}\int_0^T I(t)\mathrm{d}t \\ &= \frac{Vdd \times C}{T} \times (1 - e^{-T/RC}). \end{aligned} \tag{6.10}$$

Therefore, in order to minimize the current consumption, a combination of small C and large R is preferred. A trade-off between R, C, and α has be be found, with a compromise of area, device matching, and design feasibility. In this work, the oscillator frequency is targeted at 12.6 MHz. *Vref* is chosen at around 750 mV, which simplifies the design of the input differential pair of the comparator, and at the same time, a large RC requirement is still retained for a lower power consumption. C is chosen as 0.85 pF, which yields a fast charging slope but still a good matching between the two timing capacitors. Finally, R can be calculated from Eq. (6.4).

The main threshold-detecting comparator used in this design employs a simple two-stage open-loop structure. It has a relatively high gain and large delay. Meanwhile, the area and power consumption are kept low. The specs for the design of this comparator are quite relaxed due to the implementation of the error feedback. In order to effectively and accurately integrate the error information on the capacitor within 40 ns scale (< 1/2 *Tosc*), the amplifier used in the SC needs moderate speed and high gain. A normal class-A amplifier will consume a large amount of power to fulfill all requirements. In this design, a class-AB structure is adopted, which offers large driving ability while the static power consumption can be kept low. The amplifier has a differential input and single-ended output structure, as shown in Fig. 6.14. Chopping has also been implemented in the amplifier to reduce its offset and flicker noise, which adds to the total phase noise of the oscillator clock. The amplifier has a DC gain of 80 dB, and a static current consumption of only 20 μA.

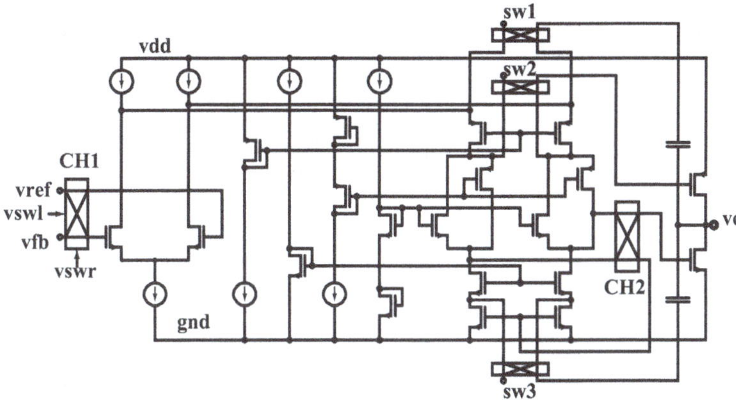

Fig. 6.14 Schematic of the differential-input single-ended output chopped class-AB amplifier implemented in the switched capacitor (*SC*) integrator

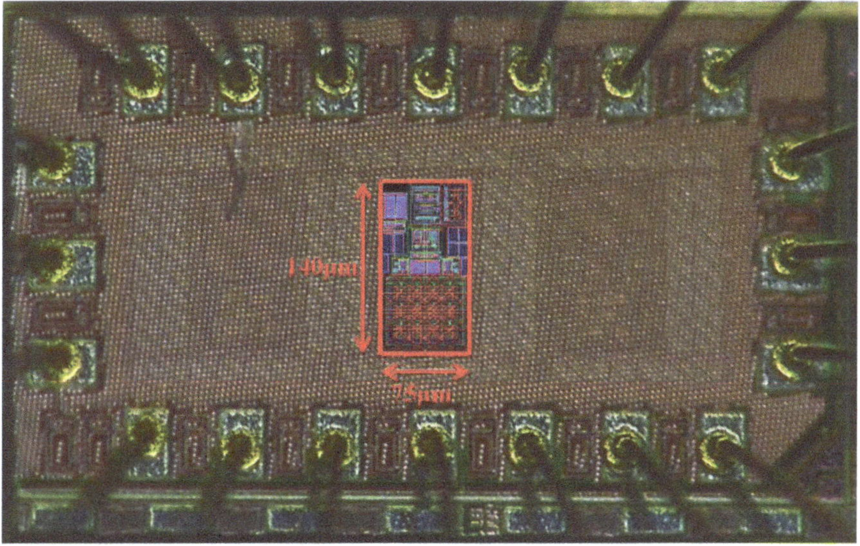

Fig. 6.15 Die photo of the relaxation oscillator with switched-capacitor integrated error feedback (*SC IEF*)

6.5.3 Experiment Results

The relaxation oscillator has been implemented in a 65 nm CMOS technology. The die photo is shown in Fig. 6.15. For the selected $R = 46$ kΩ, $C = 0.85$ pF, $R1 = 45$ kΩ, $R2 = 30$ kΩ, and $R3 = 50$ kΩ, the oscillator frequency is $f_{osc} = 12.6$ MHz.

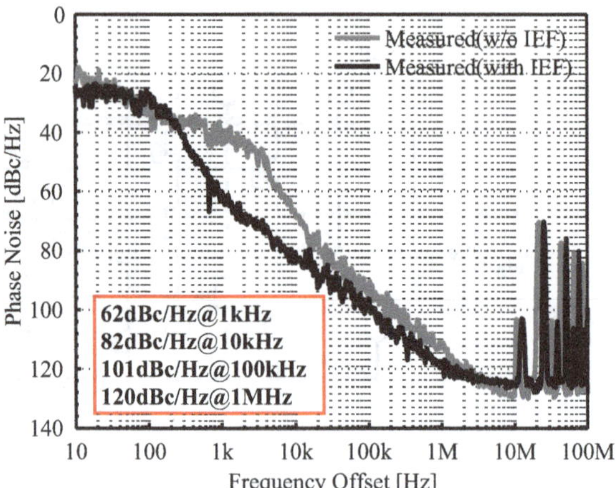

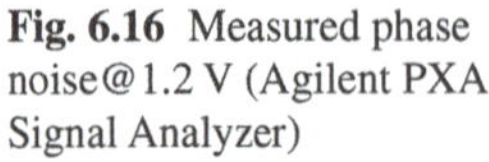
Fig. 6.16 Measured phase noise@1.2 V (Agilent PXA Signal Analyzer)

The oscillator core, including the IEF, occupies an area of only 0.01 mm^2, and consumes 82 μA from a 1.2 V supply.

Figure 6.16 shows the measured phase noise performance of the oscillator at 1.2 V. It can be clearly seen that, the low-frequency noise has been greatly suppressed by the IEF. In comparison of the phase noise spectrum with and without IEF, the crossover frequency between flicker and thermal noises has been reduced from 100 to 1 kHz by enabling IEF, which resulted in a significant reduction of the phase noise at low-offset frequency. This enables the oscillator to achieve a phase noise of -62 dBc/Hz at 1 kHz offset frequency and -101 dBc/Hz at 100 kHz offset frequency. At a large offset frequency of 1 MHz, the phase noise is -120 dBc/Hz. The -20 dB/dec declining slope is clearly obtained between 1 kHz and 1 MHz.

A commonly used figure of merit (FOM) [32] for comparing the phase noise performance of different oscillators is given by

$$\begin{aligned} \text{FOM} &= 10 \cdot \log\left(\frac{f_{osc}^2}{f_m^2 \cdot L(f_m)} \times \frac{1}{P_{diss}}\right) \\ &= 10 \cdot \log\left(\frac{T_{osc}}{\sigma_{T_{osc}}^2} \times \frac{1}{P_{diss}}\right), \end{aligned} \tag{6.11}$$

in which P_{diss} is the dissipated power in the core of the oscillator. $L(f_m)$ is the common measure for phase noise, and it is given by [31]

$$L(f_m) = \left(\frac{f_{osc}}{f_m^2} \times \left(\frac{\sigma_{\Delta T_{osc}}}{T_{osc}}\right)^2\right), \tag{6.12}$$

where f_m is the carrier offset frequency and $(\sigma_{\Delta T_{osc}}/T_{osc})$ is the ppm rms jitter. Measured FOM of the relaxation oscillator with SC IEF were 154.1 and 152.6 dB at 1 and 100 kHz offset, respectively.

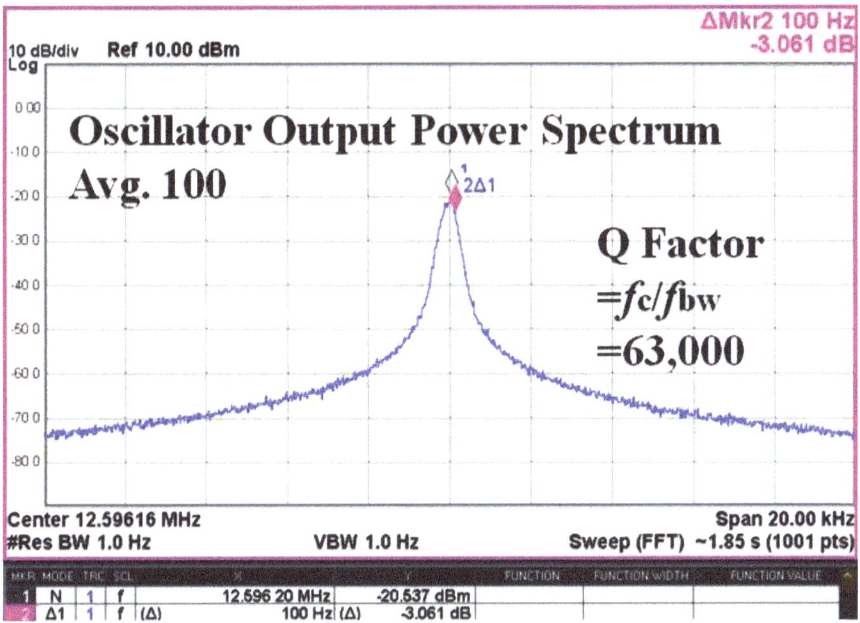

Fig. 6.17 Measured quality factor of the relaxation oscillator with switched-capacitor integrated error feedback (*SC IEF*)

The reduction of the flicker noise in the relaxation oscillator will lead to a high quality factor and small accumulated jitter. The quality factor of an oscillator is defined as

$$Q = \frac{f_c}{\Delta f}, \tag{6.13}$$

in which f_c is the oscillator center frequency, and Δf is the half-power bandwidth. A higher quality factor represents that more energy is concentrated at the oscillator center frequency. In the power spectrum (see Fig. 6.17) of the test oscillator's output clock, the −3 dB bandwidth is found to be only 200 Hz. This results in a quality factor of 63,000, which is the best reported in state of the art (see Table 6.1).

Figure 6.18 shows that almost the same period jitter were measured with and without IEF. However, suppression of close-in phase noise with IEF results in a significant reduction of accumulated jitter from 30 to 1 ns at 1000th cycle. The measured accumulated jitter with IEF is proportional to the square root of the number of periods. As discussed in Sect. 6.3.2, this is the case only when the long-term jitter is dominated by uncorrelated noise, which confirms the removal of $1/f$ noise.

Besides the frequency stability of the oscillator clock, its absolute accuracy is also of importance. The accuracy of the oscillator clock is mainly subjected to the process, supply voltage, and temperature variation. Without any tuning, a frequency spread

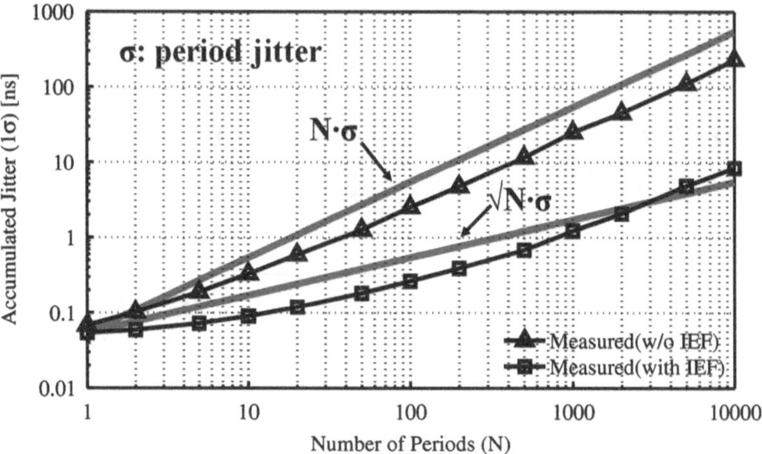

Fig. 6.18 Measured accumulated jitter

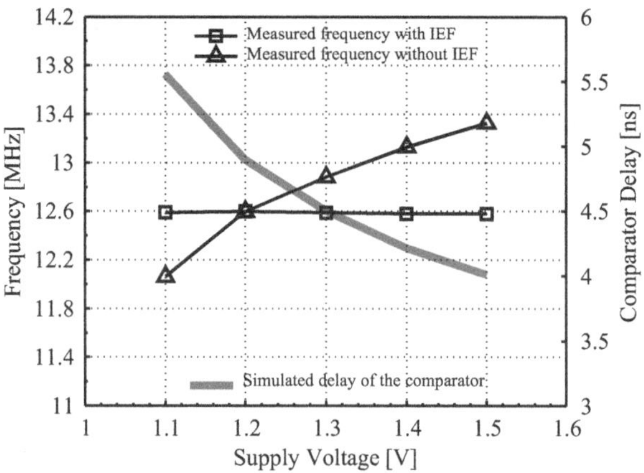

Fig. 6.19 Measured oscillator frequency versus power supply

of ± 1.6 % has been found from ten randomly picked measured samples, which is mainly due to the process variation.

Figure 6.19 shows the measured frequency versus supply voltage. According to Eq. (6.4), the oscillation period should be independent of the supply voltage. However, without IEF, the delay of the comparator (two times) will also become a consistent part of the clock period. And the comparator delay has a strong dependence on the supply voltage. Simulation results have shown that the comparator has less delay when it is working under higher supply voltage. Without the SC IEF loop, this will result in a frequency shift as much as nearly 11 % (oscillation frequencies at 1.2 V supply have been first tuned to be equal for both oscillators with and without IEF). By applying the SC IEF, the comparator delay variation can be compensated, and the clock frequency is able to remain stable over the entire supply range. As

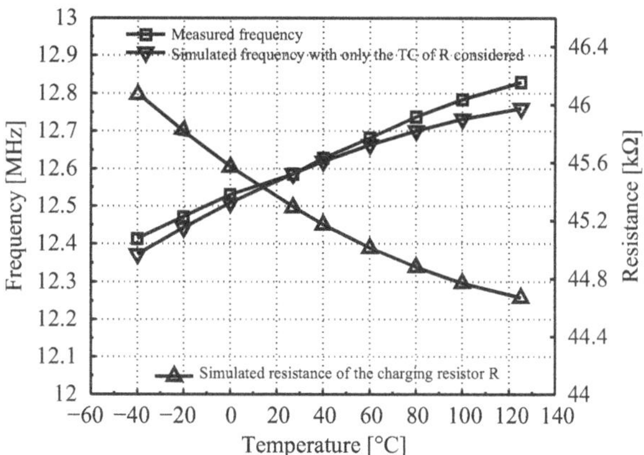

Fig. 6.20 Measured oscillator frequency versus temperature

clearly seen in Fig. 6.19, the measured frequency change of the oscillator subjected to supply voltage variation is only ±0.07 % over a range of 1.1–1.5 V, which proves the effectiveness of the SC IEF loop.

The oscillator has also been measured under different temperatures. The temperature stability of the oscillator is mainly determined by the thermal characteristic of the charging resistor *R*, which is shown in the right axis of Fig. 6.20. The measured frequency drift over temperature is very close to the simulated value. From 0 to 80 °C, the total frequency variation is within ± 0.82 %. This could be further calibrated by applying an appropriate temperature gradient to vref according to the temperature coefficient of the charging resistor.

Table 6.1 presents the performance summary and comparison with other state-of-the-art works which have similar specifications and also with some recent MEMS oscillators. The proposed oscillator achieves the best-reported quality factor in the category of relaxation oscillators with minimum area occupation. MEMS oscillators indeed have better frequency stability and lower close-in phase noise. However, this comes at the cost of a higher power consumption, larger area, and the requirement of special process steps. The relaxation oscillator with SC IEF proves a solution for high-Q low-cost on-chip clock generation.

6.6 Conclusions

A low-power area-efficiency low-jitter relaxation oscillator has been introduced in this chapter. It is intended to be implemented in the MASH $\Delta\Sigma$ TDC for achieving better time resolution, but it can also be used as a primary clock source for low-cost SoCs.

Compared to the XO, MEMS oscillator, and LC oscillator, the relaxation oscillator offers a very attractive approach to generate a reference clock for low-spec SoCs, due to their low cost, high integration level, good tuning linearity, and area efficiency.

Table 6.1 Performance comparison

	[30]	[87]	[88]	[52]	[85]	This work
Feature	Relaxation	Relaxation	Relaxation	MEMS	MEMS	Relaxation
Technology	65 nm	0.18 μm	65 nm	0.35 μm	MEMS: HARPSS-on-SOI	65 nm
	CMOS	CMOS	CMOS	CMOS	IC: 0.5 μm CMOS	CMOS
Area (mm^2)	0.03	0.04	0.02	0.026	2.25	0.01
Frequency (MHz)	12	14	6	60	5.5	12.6
Vdd (V)	1.3	1.8	1.25	MEMS: 12	MEMS: 3–15	1.2
				IC: $\pm$1.65	IC: 5	
Power cons. (μW)	91	43.2	100	950	1800	98.4
Q-Factor	N/A	11,666	23,885	48,000	Resonator: 103,000	63,000
					Oscillator: 54,000	
FOM (dB)	162	146	117.6	N/A	N/A	154.1@1 kHz
	@100 kHz	@100 kHz	@100 kHz			152.6@100 kHz
Variation (%)	N/A	$\pm$0.16	$\pm$0.26	llN/A	N/A	$\pm$0.07
with Vdd		@1.7–1.9 V	@1.15–1.35 V			@1.1–1.5 V
Variation (%)	N/A	$\pm$0.75	$\pm$0.2	35 ppm	332 ppm	$\pm$0.82
with temperature		@−40 to 125 °C	@−40 to 125 °C	@0–70°C	@25–125 °C	@0–80 °C

MEMS microelectromechanical system, *CMOS* complementary metal-oxide–semiconductor, *HARPSS* high aspect-ratio-combined poly- and single-crystal silicon, *SOI* silicon on insulator, *IC* integrated circuits

However, their poor phase noise and accuracy remain as major obstacles. After gaining an insight of the noise sources in the relaxation oscillator, a SC IEF technique has been proposed to solve those issues.

The suggested approach has been proved experimentally as a very effective way to improve both the oscillator's phase noise performance and clock accuracy. The tested relaxation oscillator with SC IEF demonstrates a phase noise of −101 dBc/Hz at 100 kHz offset frequency, when the center frequency is 12.6 MHz and achieves an FOM of 152.6 dB. The clock accuracy against power supply and temperature variation is ± 0.07 % (1.1–1.5 V) and ± 0.82 % (0–80 °C), respectively. These figures are good enough for this application; however, they could be further improved by applying calibration against PVT variations when there is a more stringent requirement. Overall, the relaxation oscillator with SC IEF provides a low-cost highly integrated solution for high-Q low-power on-chip clock generation.

Chapter 7
Conclusions

Radiation-tolerant integrated circuits are highly desired in high-energy particle physics experimental machines and nuclear and space instruments due to the increasing complexity in functionality and ever-demanding reliability of these systems. A common requirement for microelectronics used in a radiation environment that suffer from various radiation effects, such as total ionizing dose (TID) effects and single event effects (SEEs), is the radiation hardness assurance.

The main goal of this research is to design a megagray radiation-tolerant *ps*-resolution time-to-digital converter (TDC) for a light detection and ranging (LIDAR) system used in the multipurpose hybrid research reactor for high-tech applications (MYRRHA) reactor. Several radiation hardened by design (RHBD) techniques are demonstrated throughout the design of the TDC, e.g.: (1) system-level RHBD by employing an innovative multistage noise-shaping (MASH) TDC structure; (2) circuit-level RHBD by improving the radiation hardness of a bandgap reference implemented in the TDC system using a dynamic base leakage current compensation technique. Other circuit techniques to improve the TDC's resolution in a harsh environment are also investigated, such as the delay-line assisted calibration technique for correcting phase errors in the MASH TDC, and the integrated error feedback technique for enhancing the jitter performance of the oscillator which is the core component of the proposed TDC system.

- Commonly used TDC structures are reviewed in this book. By carefully analyzing both their advantages and performance constraints, it has been concluded that in order to design a megagray radiation-tolerant TDC, hierarchical RHBD strategies including system-level, circuit-level, device-level, and layout-level approaches must be adopted. An innovative MASH TDC architecture is therefore proposed to fulfill both the megagray radiation tolerance and high-resolution requirements.
- The first $\Delta\Sigma$ TDC which successfully implemented the third-order noise-shaping in time domain is introduced. It brings us a new type of TDC, whose resolution is not limited by the intrinsic complementary metal–oxide–semiconductor (CMOS) gate delay and is not sensitive to the process, voltage, temperature (PVT) variation and radiation effects. The demonstrated 1-1-1 MASH $\Delta\Sigma$ TDC is implemented in 0.13 μm CMOS and achieves a time resolution of 5.6 ps, when the oversampling ratio OSR is 250. It consumes 1.7 mW from a 1.2 V supply and exhibits an effective

© Springer International Publishing Switzerland 2015

Y. Cao et al., *Radiation-Tolerant Delta-Sigma Time-to-Digital Converters*,
Analog Circuits and Signal Processing, DOI 10.1007/978-3-319-11842-0_7

number of bit (ENOB) of 11 bits. The time resolution of the TDC is mainly limited by the comparator delay, which causes phase skew error when turning on/off the TDC. By employing a delay-line-assisted calibration technique, this skew error can be mostly corrected. The calibrated MASH $\Delta\Sigma$ TDC has further reduced power consumption of only 0.7 mW and achieves an ENOB of 13 bits. A wide input range of 100 ns has also been obtained. Owing to the usage of circuit level radiation-hardened-by-design techniques, such as passive RC oscillators and constant-g_m biasing, the TDC exhibits enhanced radiation tolerance. At a low dose rate of 1.2 kGy/h, the frequency of the counting clock in the MASH TDC remains constant up to at least 160 kGy. Even after a total dose of 3.4 MGy at a high dose rate of 30 kGy/h, the TDC still achieves a time resolution of 10.5 ps.

- A bandgap voltage reference plays a critical role in the MASH $\Delta\Sigma$ TDC by providing the reference voltage/current for the system. It is required to be stabilized over PVT variations and accumulated ionizing radiation dose. The TID tolerance of bandgap references in deep-submicron CMOS technology is generally limited by the radiation introduced leakage current in diodes. In this work, a dynamic base leakage compensation (DBLC) technique is proposed to eliminate this leakage current, and improving the radiation hardness of a CMOS bandgap reference to a megagray level. The silicon-proven radiation-tolerant bandgap reference exhibits a temperature coefficient of 15 ppm/°C from −40 to 125 °C, and the voltage variation from 0 to 100 °C is only ± 1 mV. The output voltage of the bandgap reference using the DBLC technique has shown a variation of only 3 % after 4.5 MGy during a real-time gamma irradiation assessment with a dose rate of 27 kGy/h.
- The best achievable time resolution of the proposed MASH $\Delta\Sigma$ TDC is highly relying on the employed relaxation oscillator's phase noise performance. Besides this application, relaxation oscillators can also been used as voltage-controlled oscillators (VCOs) in phase-locked loop (PLL) circuitries and on-chip reference clock generators. However, two major problems hindering their widespread usage are the poor phase noise performance and large long-term variation. An anti-jitter technique based on the switched-capacitor (SC) integrated error feedback (IEF) is proposed to tackle these issues. The suggested approach provides a low-power area-efficient solution for improving both the oscillator's phase noise performance and clock accuracy. The tested relaxation oscillator with SC IEF demonstrates a phase noise of −101 dBc/Hz at 100 kHz offset frequency, when the center frequency is 12.6 MHz and achieves an figure of merit (FOM) of 152.6 dB. The clock accuracy against power supply and temperature variation is ± 0.07 % (1.1–1.5 V) and ± 0.82 % (0–80 °C), respectively.

References

1. Abderrahim H, Baeten P, Bruyn DD, Fernandez R (2012) MYRRHA—a multi-purpose fast spectrum research reactor. In: Proceedings of 10th International Conference on Sustainable Energy Technologies, vol 63, pp 4–10
2. Abidi AA (1987) Linearization of voltage-controlled oscillators using switched-capacitor feedback. IEEE J Solid-State Circuits 22(3):494–496
3. Abidi AA, Meyer RG (1983) Noise in relaxation oscillators. IEEE J Solid-State Circuits 18(6):794–802
4. Agarwal V, Birkar SD (2005) Comparison of gamma radiation performance of a range of CMOS A/D converters under biased conditions. IEEE Trans Nucl Sci 52(6):3059–3067
5. Amann M, Bosch T, Lescure M, Myllyla R, Rioux M (2001) Laser ranging: a critical review of usual techniques for distance measurement. Opt Eng 40(1):10–19
6. Anelli G, Campbell M, Delmastro M et al (1999) Radiation tolerant VLSI circuits in standard deep submicron CMOS technologies for the LHC experiments: practical design aspects. IEEE Trans Nucl Sci 46(6):1690–1696
7. Armstrong SE, Olson BD, Holman WT et al (2010) Demonstration of a differential layout solution for improved ASET tolerance in CMOS A/MS circuits. IEEE Trans Nucl Sci 57(6):3615–3619
8. Banba H, Shiga H, Umezawaand A et al (1999) A CMOS bandgap reference circuit with sub-1-V operation. IEEE J Solid-State Circuits 34(5):670–674
9. Barnaby HJ (2006) Total-ionizing-dose effects in modern CMOS technologies. IEEE Trans Nucl Sci 53(6):3103–3121
10. Barnett R, Liu J (2006) A 0.8V 1.52 MHz MSVC relaxation oscillator with inverted mirror feedback reference for UHF RFID. In: Proceedings Custom Integrated Circuits Conference, pp 769–772
11. Baumann R (2005) Single-event effects in advanced CMOS technology. In: IEEE Nuclear and Space Radiation Effects Conference Short Course Notebook
12. Bazes M (1991) Two novel fully complementary self-biased CMOS differential amplifiers. IEEE J Solid-State Circuits 26(2):165–168
13. Beheim G, Fritsch K (1986) Range finding using frequency-modulated laser diode. Appl Opt 25(9):1439–1442
14. Blaine RW, Armstrong SE, Kauppila JS et al (2011) RHBD bias circuits utilizing sensitive node active charge cancellation. IEEE Trans Nucl Sci 58(6):3060–3066
15. Bosch T, Leacure M (1995) Selected papers on laser distance measurements. SPIE Press, Washington
16. Cao Y, Leroux P, Cock WD, Steyaert M (2011a) A 0.7mW 13b temperature-stable MASH $\Delta\Sigma$ TDC with delay-line assisted calibration. In: Proceedings of the IEEE Asian Solid-State Circuits Conference, pp 361–364
17. Cao Y, Leroux P, Cock WD, Steyaert M (2011b) A 1.7 mW 11b 1-1-1 MASH $\Delta\Sigma$ time-to-digital converter. In: IEEE International Solid-State Circuits Conference, Digest of Technical Papers, pp 480–481

© Springer International Publishing Switzerland 2015
Y. Cao et.al., *Radiation-Tolerant Delta-Sigma Time-to-Digital Converters*,
Analog Circuits and Signal Processing, DOI 10.1007/978-3-319-11842-0

18. Casey MC, Armstrong SE, Arora R (2009) Effect of total ionizing dose on a bulk 130 nm ring oscillator operating at ultra-low power. IEEE Trans Nucl Sci 56(6):3262–3266
19. CERN (2013) ATLAS experiment website. http://www.atlas.ch/pixel-detector.html. Accessed 25 March 2013
20. Cester A, Paccagnella A (2004) Switching oxide ionizating radiation effects on ultra-thin oxide MOS structures. In: Schrimpf RD, Fleetwood DM (eds) Radiation effects and soft errors in integrated circuits and electronic devices. World Scientific, Singapore
21. Choe K, Bernal OD, Nuttman D, Je M (2009) A precision relaxation oscillator with a self-clocked offset-cancellation scheme for implantable biomedical SoCs. In: IEEE International Solid-State Circuits Conference, Digest of Technical Papers, pp 402–403
22. Dodd PE, Shaneyfelt MR, Schwank JR, Felix JA (2010) Current and future challenges in radiation effects on CMOS electronics. IEEE Trans Nucl Sci 57(4):1747–1763
23. Drago S, Sebastiano F, Breems L et al (2009) Impulse-based scheme for crystal-less ULP radios. IEEE Trans Circuits Syst I 56(5):1041–1052
24. Drakhlis B (2001) Calculate oscillator jitter by using phase-noise analysis, Part 2. Microw RF 82–90, 157
25. Dudek P, Szczepanski S, Hatfield JV (2000) A high-resolution CMOS time-to-digital converter utilizing a vernier delay line. IEEE J Solid-State Circuits 35(2):240–247
26. ESA (2010) Total dose steady-state irradiation test method. ESA/SCC basic specifications No. 22900. https://escies.org/download/webDocumentFile?id=760. Accessed 10 January 2013
27. Faccio F, Cervelli G (2005) Radiation-induced edge effects in deep submicron CMOS transistors. IEEE Trans Nucl Sci 52(6):2413–2420
28. Flynn MP, Lidholm SU (1992) A 1.2-μm CMOS current-controlled oscillator. IEEE J Solid-State Circuits 27(7):982–987
29. Furano G, Jansen R, Menicucci A (2013) Review of radiation hard electronics activities at European Space Agency. J Instrum 8:1–12
30. Geraedts P, van Tuijl E, Klumperink E et al (2008) A 90 μW 12 MHz relaxation oscillator with a-162dB FOM. In: IEEE International Solid-State Circuits Conference, Digest of Technical Papers, pp 348–349
31. Gierkink SLJ (1999) Control linearity and jitter of relaxation oscillators. PhD thesis, Universiteit Twente
32. Gierkink SLJ, van Tuijl E (2002) A coupled sawtooth oscillator combining low jitter with high control linearity. IEEE J Solid-State Circuits 37(6):702–710
33. Giraud A (2004) Radiation tolerance assessment of standard electronic components for remote handling. Annual Report of the Association EURATOM-CEA. http://www-fusion-magnetique.cea.fr/actualities/RA04/eur-cea-technology-2004-full.pdf. Accessed 5 May 2009
34. Gromov V, Annema AJ, Kluit R, Visschers JL (2007) A radiation hard bandgap reference circuit in a standard 0.13 μm CMOS technology. IEEE Trans Nucl Sci 54(6):2727–2733
35. Gruss A, Carley L, Kanade T (1991) Integrated sensor and range-finding analog signal processor. IEEE J Solid-State Circuits 26(3):184–191
36. Gupta G, Mukhopadhyay S, Messom C, Demidenko S (2005) Master-slave control of a teleoperated anthropomorphic robotic arm with gripping force sensing. In: Proceedings of the IEEE Instrumentation and Measurement Technology Conference, vol 3, pp 2203–2208
37. Hashimoto T, Yamazaki H, Muramatsu A, Sato T, Inoue A (2008) Time-to-digital converter with vernier delay mismatch compensation for high resolution on-die clock jitter measurement. In: Proceedings of the IEEE Symposium on VLSI Circuits, pp 166–167
38. Henzler S, Koeppe S, Kamp W, Mulatz H, Schmitt-Landsiedel D (2008) 90 nm 4.7 ps-resolution 0.7-LSB single-shot precision and 19 pJ-per-shot local passive interpolation time-to-digital converter with on-chip characterization. In: IEEE International Solid-State Circuits Conference, Digest of Technical Papers, pp 548–549
39. IAEA (2013) Power reactor information system. http://www.iaea.org/PRIS/home.aspx. Accessed 5 April 2013
40. ITER (2013) Official website of the ITER Tokamak. http://www.iter.org. Accessed 8 April 2013

41. Jansson JP, Mäntyniemi A, Kostamovaara J (2006) A CMOS time-to-digital converter with better than 10 ps single-shot precision. IEEE J Solid-State Circuits 41(6):1286–1296
42. Johnston AH, Swimm RT, Allen GR, Miyahira TF (2009) Total dose effects in CMOS trench isolation regions. IEEE Trans Nucl Sci 56(4):1941–1949
43. Koskinen M, Kostamovaara J, Myllylae R (1991) Comparison of continuous-wave and pulsed time-of-flight laser range-finding techniques. In: Proceedings of SPIE, vol 1614, pp 296–305
44. Kujik KE (1973) A precision reference voltage source. IEEE J Solid-State Circuits 8:222–226
45. Kurtti S, Kostamovaara J (2009) Laser radar receiver channel with timing detector based on front end unipolar-to-bipolar pulse shaping. IEEE J Solid-State Circuits 44(3):835–847
46. Lacoe R (2003) CMOS scaling, design pinciples and hardening-by-design methodologies. In: IEEE Nuclear and Space Radiation Effects Conference Short Course Notebook
47. Lacoe RC (2008) Improving integrated circuit performance through the application of hardness-by-design methodology. IEEE Trans Nucl Sci 55(4):1903–1925
48. Lacoe RC, Osborn JV, Koga R et al (2000) Application of hardness-by-design methodology to radiation-tolerant ASIC technologies. IEEE Trans Nucl Sci 47(6):2334–2341
49. Lee M, Abidi AA (2008) A 9b, 1.25 ps resolution coarse-fine time-to-digital converter in 90 nm CMOS that amplifies a time residue. IEEE J Solid-State Circuits 43(4):769–777
50. Leroux P, Lens S, Voorspoels R et al (2007) Design and assessment of a circuit and layout level radiation hardened CMOS VCSEL driver. IEEE Trans Nucl Sci 54(4):1055–1060
51. Leung KN, Mok PKT (2002) A sub-1-V 15-ppm/°C CMOS bandgap voltage reference without requiring low threshold voltage device. IEEE J Solid-State Circuits 37(4):526–530
52. Lin Y, Lee S, Li S et al (2004) 60-MHz wine-glass micromechanical-disk reference oscillator. In: IEEE International Solid-State Circuits Conference, Digest of Technical Papers, pp 322–323
53. Mäntyniemi A, Rahkonen T, Kostamovaara J (2009) A CMOS time-to-digital converter (TDC) based on a cyclic time domain successive approximation interpolation method. IEEE J Solid-State Circuits 44(11):3067–3078
54. McCorquodale MS, O'Day JD, Pernia SM et al (2007) A monolithic and self-referenced RF *LC* clock generator compliant with USB 2.0. IEEE J Solid-State Circuits 42(2):385–399
55. McLean FB, Oldham TR (1987) Basic mechanisms of radiation effects in electronic materials and devices. Tech. Rep. HDL-TR-2129, Harry Diamond Laboratory. www.dtic.mil/dtic/tr/fulltext/u2/a186936.pdf. Accessed 12 March 2013
56. Moreira P (2004) Radiation effects on the "CERN_Bandgap" circuit. http://proj-qpll.web.cern.ch/proj-qpll/images/bandgapRadEffects.pdf. Accessed 15 April 2010
57. Navid R, Lee TH, Dutton RW (2005) Minimum achievable phase noise of *RC* oscillators. IEEE J Solid-State Circuits 40(3):630–637
58. Nissinen J, Nissinen I, Kostamovaara J (2009) Integrated receiver including both receiver channel and TDC for a pulsed time-of-flight laser rangefinder with cm-level accuracy. IEEE J Solid-State Circuits 44(5):1486–1497
59. Norsworthy SR, Schreier R, Temes GC (1997) Delta-sigma data converters: theory, design, and simulation. IEEE Press, New York
60. Oldham TR, Mclean FB (2003) Total ionizing dose effects in MOS oxides and devices. IEEE Trans Nucl Sci 50(3):483–499
61. Palojärvi P, Ruotsalainen T, Kostamovaara J (2005) A 250-MHz BiCMOS receiver channel with leading edge timing discriminator for a pulsed time-of-flight laser rangefinder. IEEE J Solid-State Circuits 40(6):1341–1349
62. Pater J (2012) The ATLAS semiconductor tracker operation and performance. J Instrum 7:1–12
63. Pease RL (2003) Total ionizing dose effects in bipolar devices and circuits. IEEE Trans Nucl Sci 50(3):539–551
64. Perrott MH, Salvia JC, Lee FS et al (2013) A temperature-to-digital converter for a MEMS-based programmable oscillator with $< \pm 0.5$-ppm frequency stability and < 1-ps integrated jitter. IEEE J Solid-State Circuits 48(1):276–291
65. Poujouly S, Journet B (2001) Laser range-finding by phase-shift measurement: moving toward smart systems. In: Proceedings of SPIE, vol 4189, pp 152–160

66. Ratti L, Gaioni L, Manghisoni M et al (2008) Investigating degradation mechanisms in 130 nm and 90 nm commercial CMOS technologies under extreme radiation conditions. IEEE Trans Nucl Sci 55(4):1992–2000
67. Razavi B (1996) A study of phase noise in CMOS oscillators. IEEE J Solid-State Circuits 31(4):331–343
68. Ribeiro I, Damiani C, Tesini A, Kakudate S, Siuko M, Neri C (2011) The remote handling systems for ITER. In: Proceedings of 26th Symposium of Fusion Technology, vol 86, pp 471–477
69. Ruotsalainen ER, Rahkonen T, Kostamovaara J (2000) An integrated time-to-digital converter with 30-ps single-shot precision. IEEE J Solid-State Circuits 35(10):1507–1510
70. Ruotsalainen T, Palojärvi P, Kostamovaara J (2001) A wide dynamic range receiver channel for a pulsed time-of-flight laser radar. IEEE J Solid-State Circuits 36(8):1228–1238
71. Schwank JR, Shaneyfelt MR, Fleetwood DM et al (2008) Radiation effects in MOS oxides. IEEE Trans Nucl Sci 55(4):1833–1853
72. SCK (2013) Official website of the MYRRHA project. http://myrrha.sckcen.be/en. Accessed 9 April 2013
73. Sebastiano F, Breems LJ, Makinwa KAA et al (2009) A low-voltage mobility-based frequency reference for crystal-less ULP radios. IEEE J Solid-State Circuits 44(7):2002–2009
74. Seo YH, Kim JS, Park HJ, Sim JY (2011) A 0.63 ps resolution, 11b pipeline TDC in 0.13 μm CMOS. In: Proceedings of the IEEE Symposium on VLSI Circuits, pp 152–153
75. Shakiba MH, Sowlati T (1998) Automatic swing control in relaxation oscillators. IEEE J Solid-State Circuits 33(12):1979–1986
76. Shaneyfelt MR, Schwank JR, Fleetwood DM et al (1992) Interface-trap buildup rates in wet and dry oxides. IEEE Trans Nucl Sci 39(6):2244–2251
77. Shaneyfelt MR, Fleetwood DM, Winokur PS et al (1993) Effects of device scaling and goemetry on MOS radiation hardness assurance. IEEE Trans Nucl Sci 40(6):1678–1685
78. Shaneyfelt MR, Schwank JR, Dodd PE, Felix JA (2008) Total ionizing dose and single event effects hardness assurance qualification issues for microelectronics. IEEE Trans Nucl Sci 55(4):1926–1946
79. Shinohara S, Yoshida H, Ikeda H et al (1992) Compact and high-precision range finder with wide dynamic range and its application. IEEE Trans Instrum Meas 41(1):40–44
80. Snoeys W, Anelli G, Campbell M et al (2000) Integrated circuits for particle physics experiments. IEEE J Solid-State Circuits 35(12):2018–2030
81. Snoeys WJ, Gutierrez TAP, Anelli G (2002) A new NMOS layout structure for radiation tolerance. IEEE Trans Nucl Sci 49(4):1829–1833
82. Staszewski RB, Vemulapalli S, Vallur P, Wallberg J, Balsara PT (2006) 1.3 V 20 ps time-to-digital converter for frequency synthesis in 90-nm CMOS. IEEE Trans Circuits Syst I 53(3):220–224
83. Straayer MZ, Perrott MH (2009) A multi-path gated ring oscillator TDC with first-order noise shaping. IEEE J Solid-State Circuits 44(4):1089–1098
84. Sundaresan K, Allen PE, Ayazi F (2006) Process and temperature compensation in a 7-MHz CMOS clock oscillator. IEEE J Solid-State Circuits 41(2):433–442
85. Sundaresan K, Ho G, Pourkamali S et al (2007) Electronically temperature compensated silicon bulk acoustic resonator reference oscillators. IEEE J Solid-State Circuits 42(6):1425–1434
86. Swann BK, Blalock BJ, Clonts LG (2004) A 100-ps time-resolution CMOS time-to-digital converter for positron emission tomography imaging applications. IEEE J Solid-State Circuits 39(11):1839–1852
87. Tokunaga Y, Sakiyama S, Matsumoto A et al (2009) An on-chip CMOS relaxation oscillator with power averaging feedback using a reference proportional to supply voltage. In: IEEE International Solid-State Circuits Conference, Digest of Technical Papers, pp 404–405
88. Tokunaga Y, Sakiyama S, Dosho S (2010) An over 20,000 quality factor on-chip relaxation oscillator using power averaging feedback with a chopped amplifier. In: Proceedings of the IEEE Symposium on VLSI Circuits, pp 111–112

89. Underhill MJ (2001) The adiabatic anti-jitter circuit. IEEE Trans Ultrason Ferroelectr Freq Control 48(3):666–674
90. Verbeeck J, Leroux P, Steyaert M (2011) Design and assessment of a robust voltage amplifier with 2.5 GHz GBW and > 100 kGy total dose tolerance. J Instrum 6:1–7
91. Vroonhoven CPLV, Makinwa KAA (2008) A CMOS temperature-to-digital converter with an inaccuracy of $\pm 0.5\,°C$ (3σ) from -55 to 125°C. In: IEEE International Solid-State Circuits Conference, Digest of Technical Papers, pp 576–577
92. Wakayama MH, Abidi AA (1987) A 30-MHz low-iitter high-linearity CMOS voltage-controlled oscillator. IEEE J Solid-State Circuits 22(6):1074–1081
93. Wong A, Kathiresan G, Eljamaly O et al (2008) A 1 V wireless transceiver for an ultra-low-power SoC for biotelemetry applications. IEEE J Solid-State Circuits 43(7):1511–1521
94. Young B, Kwon S, Elshazly A, Hanumolu PK (2010) A 2.4 ps resolution 2.1 mW second-order noise-shaped time-to-digital converter with 3.2 ns range in 1 MHz bandwidth. In: Proceedings Custom Integrated Circuits Conference
95. Yu J, Dai FF, Jaeger RC (2010) A 12-bit Vernier ring time-to-digital converter in 0.13 μm CMOS technology. IEEE J Solid-State Circuits 45(4):830–842

Index

© Springer International Publishing Switzerland 2015

Y. Cao et.al., *Radiation-Tolerant Delta-Sigma Time-to-Digital Converters*,
Analog Circuits and Signal Processing, DOI 10.1007/978-3-319-11842-0